Deformation and Fracture Mechanics of Engineering Materials

Deformation and Fracture Mechanics of Engineering Materials

Editor

Deepak Gupta

Deformation and Fracture Mechanics of Engineering Materials

Edited by **Deepak Gupta**

Printed in 2017

ISBN: 978-1-68117-214-9

Library of Congress Control Number: 2015936576

© 2016 by
SCITUS Academics LLC,
616, Corporate Way, Suite 2, 4766,
Valley Cottage, NY 10989

www.scitusacademics.com

This book contains information obtained from highly regarded resources. Copyright for individual articles remains with the authors as indicated. All chapters are distributed under the terms of the Creative Commons Attribution License, which permits unrestricted use, distribution, and reproduction in any medium, provided the original author and source are credited.

Notice

Reasonable efforts have been made to publish reliable data and views articulated in the chapters are those of the individual contributors, and not necessarily those of the editors or publishers. Editors or publishers are not responsible for the accuracy of the information in the published chapters or consequences of their use. The publisher believes no responsibility for any damage or grievance to the persons or property arising out of the use of any materials, instructions, methods or thoughts in the book. The editors and the publisher have attempted to trace the copyright holders of all material reproduced in this publication and apologize to copyright holders if permission has not been obtained. If any copyright holder has not been acknowledged, please write to us so we may rectify.

Contents

Preface vii

Chapter 1   Analysis of Fatigue and Fracture of Hot Mix Asphalt Mixtures 1
             Mohammad Jamal Khattak and Gilbert Y. Baladi

Chapter 2   Dynamic Compressive Deformation and Fracture of a Hollow
             Bulk Metallic Glass 25
             Haiyan Ye, Tao Zhang, Zhihua Wang, Hefeng Zhou, Shengbo Sang,
             Huijun Yang and Junwei Qiao

Chapter 3   High Sensitive Methods for Health Monitoring of Compressor
             Blades and Fatigue Detection 41
             Mirosław Witoś

Chapter 4   A Review on Recent Contribution of Meshfree Methods to
             Structure and Fracture Mechanics Applications 115
             S. D. Daxini and J. M. Prajapati

Chapter 5   Fracture Mechanics of Polymer Mortar Made with Recycled Raw
             Materials 151
             Marco Antonio Godoy Jurumenha; João Marciano Laredo dos Reis

Chapter 6   In situ Experimental Mechanics of Nanomaterials at the Atomic
             Scale 165
             Lihua Wang, Ze Zhang, and Xiaodong Han

Chapter 7   Rate Sensitive Continuum Damage Models and Mesh
             Dependence in Finite Element Analyses 197
             Goran Ljustina, Martin Fagerström, and Ragnar Larsson

Chapter 8   Nano-scale Machining of PolycrystallineCoppers -Effects of Grain
             Size and Machining Parameters 221
             Jing Shi, Yachao Wang, and Xiaoping Yang

Chapter 9	Hydraulic Fracturing and its Peculiarities	259
	Stefano Secchi and Bernhard A Schrefler	
	Citations	291
	Index	295

Preface

Mechanics is the body of knowledge that deals with the relationships between forces and the motion of points through space, including the material space. Material science is the body of knowledge that deals with the properties of materials, including their mechanical properties. Mechanics is very deductive—having defined some variables and given some basic premises, one can logically deduce relationships between the variables. Material science is very empirical—having defined some variables one establishes the relationships between the variables experimentally. Mechanics of materials synthesizes the empirical relationships of materials into the logical framework of mechanics, to produce formulas for use in the design of structures and other solid bodies.

Editor

Chapter 1

Analysis of Fatigue and Fracture of Hot Mix Asphalt Mixtures

Mohammad Jamal Khattak[1] and Gilbert Y. Baladi[2]

[1]Department of Civil Engineering, University of Louisiana at Lafayette, Lafayette, LA 70504-2291, USA

[2]Pavement Research Center of Excellence, Department of Civil and Environmental Engineering, Michigan State University, 3546 Engineering Building, East Lansing, MI 48824, USA

ABSTRACT

An accurate assessment of the fatigue life of hot mix asphalt (HMA) mixtures depends on the criteria used in the fatigue analysis. In the past, various studies have been conducted on crack initiation and crack propagation of the HMA mixtures. Most of these studies were focused on the beam samples with or without a sawed crack at the bottom.

This paper presents and discusses two different fatigue life criteria for two-dimensional problems represented by cylindrical samples. One criterion is based on the rate of accumulation of the tensile horizontal plastic deformation (HPD) as a function of the number of load repetitions. The second criterion is based on fracture mechanics, stress intensity factor, and the rate of crack growth with respect to the number of load repetitions. It was found that, because of three-dimensional nature of the crack growth in cylindrical samples, the Paris' law was violated. It is shown that the rate of crack growth criterion provides higher values of fatigue life relative to the rate of accumulation of HPD criterion. Although a trend could be established among the fatigue lives obtained by using the two criteria, it was found that the fatigue lives obtained from the rate of accumulation of HPD were consistent and based on the actual measurement of HPD for HMA mixtures.

INTRODUCTION

The prediction of fatigue life of hot mix asphalt mixtures (HMA) is an important aspect of pavement design. Fatigue cracks are caused by repeated traffic loading and are typically initiated at the bottom of the HMA layer where the tensile stress and strain are the highest. With increasing number of load application, the cracks propagate to the surface where they appear as one or more longitudinal cracks, which will be connected by transverse cracking to form a pattern similar to an alligator hide. Many factors affect the fatigue life of HMA pavement such as the tensile strength of the asphalt binder, traffic load, construction practices, aggregate angularity and gradation, relative stiffness of the AC, and the base material and environmental conditions such as temperature and moisture.

In the past, many efforts have been made to estimate the fatigue life of laboratory compacted HMA mixtures. Such estimates are highly dependent on the criterion used. Hence, various criteria were developed and are reported in the literature [1–12]. Monismith and Deacon (1969) [6] and Pell and Cooper (1975) [7] conducted displacement-controlled trapezoidal fatigue test and proposed that for HMA mixtures the fatigue failure of the mixture is reached when the load value drops to a half of the initial load value. On the other hand for a stress-controlled fatigue test, the failure (i.e., fatigue life) was reached when the displacement value doubles the initial displacement value [11]. The damage failure criterion was used based on the analysis of the stiffness modulus values along the number of load cycles. It was observed that the stiffness modulus decreases rapidly as the load application progresses. After this, a linear decrease of the stiffness modulus was reported. Finally, a rapid loss of stiffness modulus occurred which was related to crack development [9–11]. Hence, the fatigue life was defined as the number of load cycles at which the rapid increase in stiffness modulus occurred. Similarly, for a displacement or stress controlled test, the change in the displacement or stress along the repeated loading was plotted. Based on this criterion, the HMA specimen failed when the displacement or stress values exhibited an accelerated increase while the loading process progressed [8, 11, 12].

In this paper, a comparison between two different fatigue life criteria for two-dimensional problems represented by cylindrical samples is discussed and presented. One criterion is based on the rate of accumulation of the tensile horizontal plastic deformation (HPD) as a function of the number of load repetitions. The other criterion is based on fracture mechanics and takes into account stress intensity factor and the rate of crack growth with respect to the number of load repetitions.

MATERIALS AND METHODS

Different types of HMA mixtures were constructed tested during this study. These include the following.

AC5 and AC10 Straight and Polymer-Modified Asphalt (PMA) HMA Mixtures

The mix design for these mixtures was based on the standard Marshall Mix design procedure with 50 blows on each side of cylindrical samples. The samples were 10.16 cm in diameter and 6.35 cm thick. The Michigan Department of Transportation (MDOT) specifications of voids in mineral aggregates (VMA) and other criteria for the MDOT designated 4C HMA mixtures were followed. The polymers used for HMA modification included styrene-butadiene-styrene (SBS), styrene-ethylene-butylene-styrene (SEBS), and styrene-butadiene-rubber (SBR). Detailed information regarding the mix design and polymer-modified asphalt mixtures can be found in [5].

Mdot Hma Mixtures

These mixtures were obtained from construction sites by MDOT and were compacted at the Michigan State University laboratory. Cylindrical samples of 150 mm diameter and 76 mm thick were compacted under the pressure of 600 KPa using Gyratory compactor. MDOT specifications of VMA and other criteria for the asphalt mixtures were followed for these mixtures. be compacted samples were cut into two 35 mm thick samples using watercooled diamond saw. For some samples, a slot (25 mm × 2.5 mm) was also made at the center of the sample along the vertical diameter. Samples with a slot in the center were tested under dynamic loading.

All samples were subjected to cyclic loads at 20 and 25°C using indirect tensile cyclic load tests. First, the sample was subjected to a sustained stress of 13 KPa. When the sample came to rest, a cyclic stress of 75 KPa was applied, and the deformation of the sample was measured in three directions using linear variable differential transducers (LVDTs). Each loading cycle consisted of 0.1 second load-unload time and 0.4 second of rest period. The resilient and total moduli of the mixtures were calculated using the resilient and total deformations, respectively [4]. The sample characteristics and sample geometry are listed in Table 1.

Table 1: Sample characteristics and sample geometry for fatigue testing

Mixtures	Test temperature (°C)	Average percent air voids	MDOT specification	Diameter (mm)	Thickness (mm)
AC5 and AC10 straight mixtures	25	3.50	4C	101.6	63.5
AC5 and AC10 polymer-modified asphalt mixtures	25	3.75	4C	101.6	63.5

| MDOT AC mixture without slot | 20 | 7.50 | 4B | 150 | 35 |
| MDOT AC mixture with slot (25 mm × 2.5 mm) | 20 | 5.50 | 4B | 150 | 35 |

FATIGUE LIFE CRITERIA

The fatigue life of each test sample was analyzed using two fatigue life criteria. One criterion is based on the rate of accumulation of tensile horizontal plastic deformation (HPD) with respect to the number of load repetitions. The other is based on fracture mechanics, and it takes into account the rate of crack growth with respect to the number of load repetitions. Both criteria are discussed below.

The Rate of Accumulation of HPD Criterion

This criterion was developed at Michigan State University, and it has been used in several research studies [5]. In this criterion, it was hypothesized that, at certain number of load application, microcracks are initiated in the sample. With increasing number of load repetitions, they grow to macrocracks that can be visually detected. Once the microcracks are initiated and interconnected, the tensile horizontal plastic deformation starts to increase sharply, as a result of stress concentration at the crack tips. Therefore, in this hypothesis, the fatigue life of a test sample is defined as the number of cycles at which the rate of accumulation of HPD increases. Based on the definition, the following procedure was established for the determination of the fatigue lives of the test samples.

- From the test results, the cumulative HPD $(HPD)_N$ is plotted as a function of the number of load repetition (N). A polynomial equation is fitted to the test results such that

$$(HPD)_N = \sum_{i=1}^{N} \Delta p_i = f(N), \tag{1}$$

where p_i is the horizontal plastic deformation per load cycle.

- The rate of accumulation of HPD $(RHPD)_N$ is obtained by differentiating the above equation with respect to the number of load repetition (N) as shown below:

$$(RHPD)_N = \frac{d(HPD)_N}{d(N)} = f'(N). \tag{2}$$

- The normalized rate of accumulation of HPD $(NRHPD)_N$ at any load cycle is calculated as

$$(NRHPD)_N = \frac{(RHPD)_N}{(RHPD)_{100}}, \tag{3}$$

where $(RHPD)_{100}$ is the rate of accumulation of the HPD at 100th load cycle. (4)

- The normalized rate of accumulation of HPD is then plotted on a semilogarithmic scale as a function of the number of load repetition (N).

The slope of the curve decreases first, reaches a valley, and then starts to increase. The decrease in the normalized rate of accumulation of HPD at the beginning of the test is due to slight densification and sample seating. However when cracks are initiated, the normalized rate of accumulation of HPD also increases. Hence, in this procedure, the fatigue life is defined as the number of load repetitions at which the rate of accumulation of HPD starts to increase.

The Rate of Crack Growth Criterion

The asphalt concrete is a heterogeneous material consisting of asphalt cement and fine and coarse aggregates. The fatigue cracks initiation in composite and heterogeneous material is related to the preexisting internal flaws. In HMA mixtures these internal flaws exist in the form of air voids, surface irregularities, and internal defects. These internal flaws are randomly distributed in the HMA mixtures and control the geometry and propagation of fatigue cracks. The rate of crack growth criterion utilizes the concepts of fracture mechanics to relate the fatigue life to the rate of crack growth and stress intensity factor. With increasing number of load repetition the preexisting microcracks (internal flaws) at a given state of stress start to propagate and reach critical dimensions and pattern where the rate of crack propagation increases. The number of load repetitions at which the rate of crack growth increases is designated as the fatigue life of the HMA mixtures. The following two approaches were investigated to indirectly determine the crack length and growth.

Method of Compliance. In this fatigue life criterion an indirect method was utilized to estimate the crack length at different load repetitions. The indirect method of crack length was utilized because it is not possible to get actual measurement of the crack length in a cylindrical sample subjected to indirect tensile cyclic loading. The indirect method is based on the fact that as the crack length increases from an initial flaw, the compliance (C) of the sample also increases due to stress concentration in the vicinity of the crack tip. The compliance (C) of a material subjected to load is defined as the reciprocal of modulus and is given by the following equation [14]:

$$C = \frac{1}{E}.$$

(4)

If the material behaves elastically, the increase in compliance is related to the crack length. The increase in the compliance as a result of crack growth is expressed by Irwin's equation as follows [13, 15].

$$\frac{\partial C}{\partial a} = \frac{2}{E}\left(1 - \mu^2\right)\frac{K^2}{P^2}, \tag{5}$$

where C = compliance; E = Young's modulus; a = crack length; K = stress intensity factor, PP = load per unit thickness; μ = poisson ratio.

This equation can also be written as

$$\Delta C = \frac{2\left(1 - \mu^2\right)K^2}{E}\frac{}{P^2}\Delta a. \tag{6}$$

The stress intensity factor can be obtained as follows [16]:

$$K = \beta\sigma\sqrt{\pi a} \tag{7}$$

$$\sigma = \frac{2P}{\pi D}, \tag{8}$$

where σ is an applied stress, D is a diameter of the sample, P is the load per unit thickness, and β is a shape factor and is a function of a/D ratio as shown in Figure 1 [17].

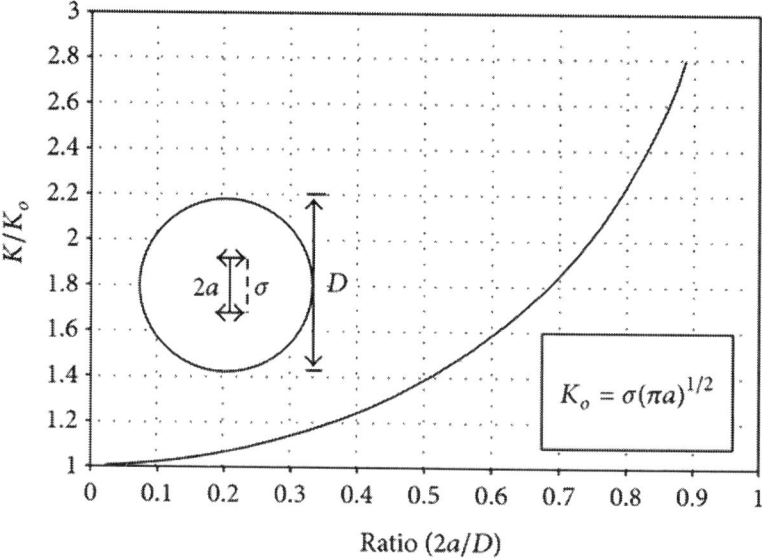

Figure 1: Stress intensity factor (K) for a centrally cracked circular plate (a er [13]).

Substituting σ in (3) and rearranging, the equation for K^2/P^2 yields the following equation:

$$\frac{K^2}{P^2} = \frac{4a\beta^2}{\pi D^2}. \tag{9}$$

If the initial compliance (C_o) and the initial crack length (a_o) are known, normalized compliance (NC) at any number of load application can be calculated from (6) as follows:

$$C - C_o = \frac{2(1-\mu^2)}{E_o} \frac{K^2}{P^2}(a - a_o),$$

$$C = C_o + \frac{2(1-\mu^2)}{E_o} \frac{K^2}{P^2}(a - a_o).\tag{10}$$

Normalizing relative to (C_o) yields

$$\frac{C}{C_o} = NC = 1 + 2(1-\mu^2)\frac{K^2}{P^2}(a - a_o).\tag{11}$$

Equation (11) gives the relation between the compliance and the crack length. From indirect cyclic load test, the compliance of a sample can be calculated at different cycles by simply taking the inverse of the total modulus (E). From this, the normalized compliance can be calculated by dividing all the data points with the compliance value at the 100th load cycle. Hence the crack length at a given load cycle can be obtained by using (11). Once the crack length is known, the stress intensity factor (K) can be calculated from (7).

The crack length as a function of number of cycles can also be calculated using another fracture mechanics technique that utilizes the vertical displacement of the sample. This technique provides a check for the crack length calculations using the compliance technique. The method of displacement is explained below.

Method of Displacement. With the increase in number of load cycles the crack length and vertical displacement of the sample increase. The vertical displacement and crack length are related to each other by the following equation [18]:

$$\frac{\Delta_p}{\Delta_p^{nocrack}} = 1 + \left[\frac{1.805P}{Et\pi\Delta_p^{nocrack}}\left(\left|\ln\left\{1 - \frac{c}{b}\right\}\right| - \frac{c}{b}\right)\right]x^2,\tag{12}$$

where Δ_p = load point displacement with crack, $\Delta_p^{nocrack}$ = load point displacement with no crack present, c = crack length, b = radius of the sample, E = elastic modulus t = thickness of the sample, and P = applied load.

The load point deflection with no crack present can be obtained from an appropriate analytical expression relating load and load point displacement to material properties and dimensions as

$$\Delta_p^{nocrack} = \frac{3.59}{Et}P. \tag{13}$$

Rearranging the above two equation will yield

$$\frac{\Delta_p}{\Delta_p^{nocrack}} = 1 + \frac{1}{\pi}\left(\left|\ln\left\{1-\frac{c}{b}\right\}\right|-\frac{c}{b}\right). \tag{14}$$

By using the numerical approach the crack length can be calculated from (14). Once the crack length is known, the K values for each load cycle can be calculated. The A and m material properties can be determined by plotting the K values and rate of crack growth in crack length on log-log scales and can be used for Paris law.

RESULTS AND DISCUSSION

Figure 2 shows the effect of the number of load repetition (N) on the modulus and normalized compliance (NC) of HMA mixture at test temperature of 20°C. It can be seen that the modulus values decrease and the normalized compliance increases with increasing number of load repetitions. A best-fit polynomial equation was obtained expressing the modulus as a function of the number of load repetition. The normalized compliance of the sample was then calculated by simply taking the inverse of the modulus and dividing the result by the compliance value at the 100th load repetition. The results were plotted as a function of the number of load repetitions and are shown in Figure 2. The normalized compliance increases as the number of cycle increases. Since the normalized compliance and the crack length are uniquely related, the crack length was calculated using (5) and plotted as a function of the number of load repetitions as shown in Figure 3. In order to validate the results the crack length was also calculated using the vertical displacement method (14) and plotted as a function of number of cycles (Figure 3). It can be seen that crack lengths calculated using both fracture mechanics methods are similar. The slight difference may be because of the best-fit curve used to fit the data. The examination of (11) and the data in Figures 2 and 3 indicate that the crack in HMA sample reaches its full length when the compliance is twice the initial value. This implies that when the modulus reaches its 50 percent value due to repeated loadings, the sample completely cracks. It should be noted that 50 percent reduction in modulus has also been used by various researchers as one of the fatigue criteria.

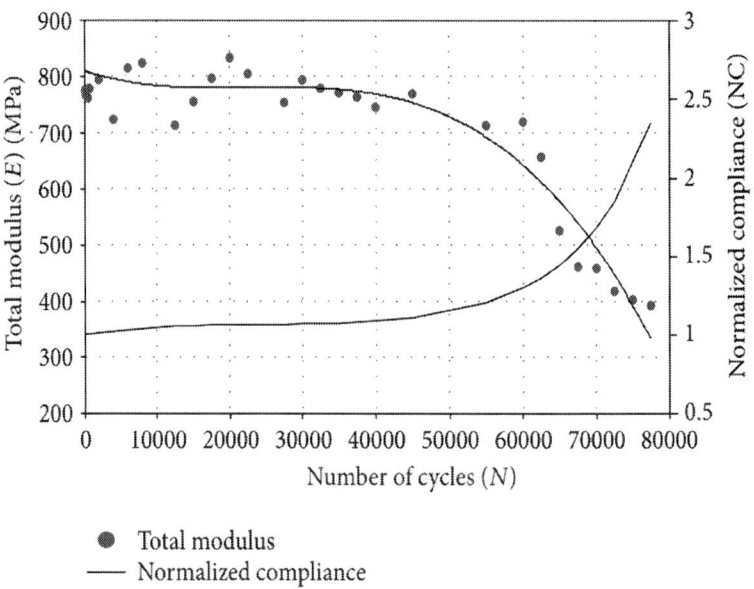

Figure 2: A typical plot of the total modulus and normalized compliance versus the number.

Analysis of Fatigue and Fracture of Hot Mix Asphalt Mixtures

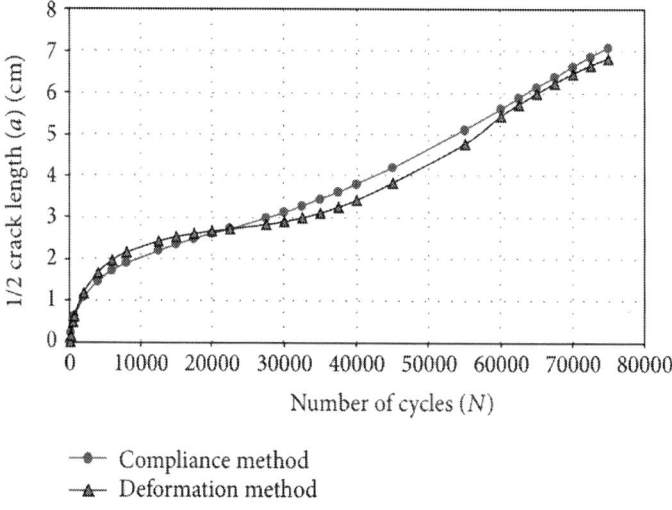

Figure 3: A typical plot of the crack length as a function of the number of load repetitions.

Similarly the measured cumulative HPD was plotted against the number of load repetition as shown in Figure4. The normalized rates of both crack growth and cumulative HPD with respect to first data point (100th load repetition) were calculated and plotted against the logarithmic value of the number of load repetition as shown in Figure 5. It can be seen that the normalized rate of cumulative HPD decreases first, forms a valley, and then starts to increase. This is due to the initial densification of the sample. Similarly, the normalized rate of crack growth decreases first, reaches a valley, and then increases. The decrease in the rate of crack growth is due to the crack being abstracted or shielded by large aggregates in the HMA mixtures. The crack had to deflect and change its path and propagate around the boundary of the aggregates. This crack deflection is referred to as crack shielding mechanism, which decreases the rate of crack growth. The crack tip shielding mechanisms have also been reported by Hertzberg [19] in polycrystalline alumina under cyclic loading. The crack growth rates were found to decrease with increasing crack length before arresting. When crack reached a critical dimension, it started to accelerate in a similar manner as that shown in Figure 5. Relative to the HMA mixtures, the number of load applications at which the crack starts to accelerate is considered the fatigue life of the HMA mixture. The stress intensity factor (K) was also calculated, and the rate of crack propagation was plotted against the stress intensity factor as shown in Figure 6. Similar results for PMA mixtures without slot and straight MDOT HMA mixtures with slot are shown in Figures 7, 8, 9, 10, and 11. These mixtures showed similar trend as that of the mixtures discussed above. The rate of crack growth decreases first and then starts to increase hence violating the Paris' law, which states that the rate of crack growth increases with the increase in the stress intensity factor and obeys the power law as given below:

$$\frac{da}{dN} = AK^m, \tag{15}$$

where A and m are material constants.

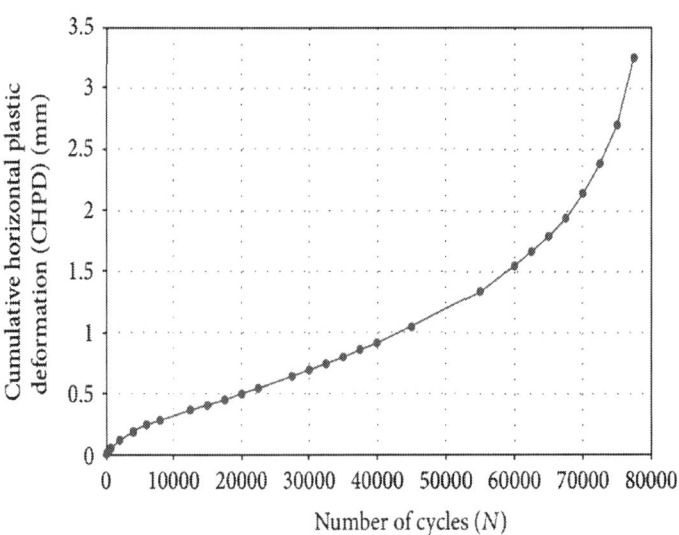

Figure 4: A typical plot of cumulative HPD as a function of the number of load repetitions.

Analysis of Fatigue and Fracture of Hot Mix Asphalt Mixtures

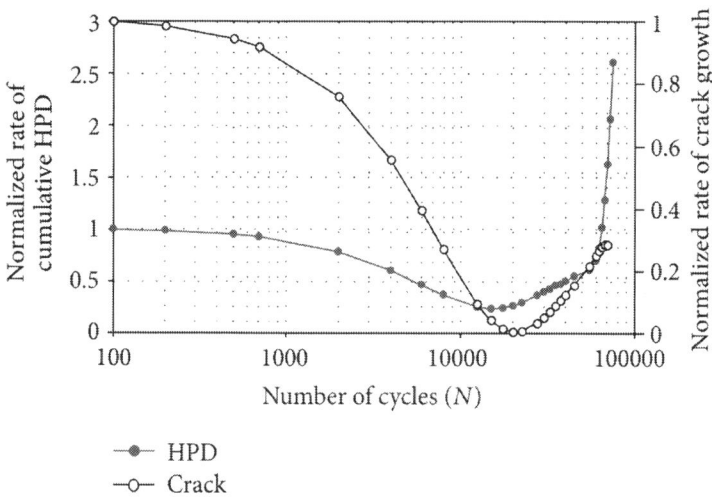

Figure 5: A typical plot of normalized rate of the rate of the crack growth and cumulative HPD as a function of the number of load repetitions.

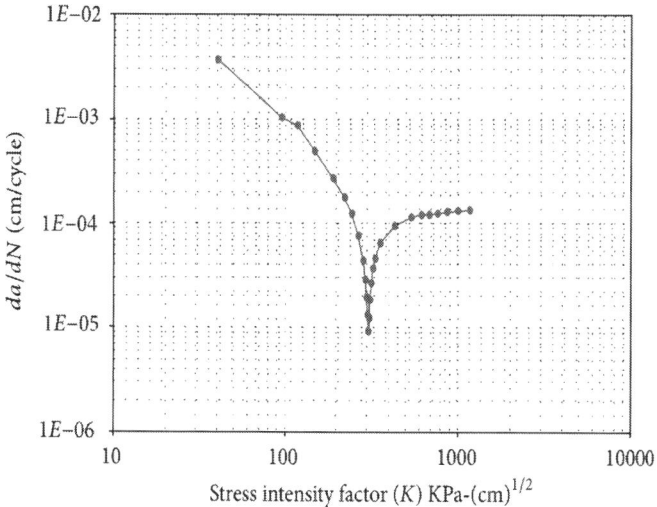

Figure 6: A typical K-da/dN curve for indirect tensile specimen of asphalt concrete mixtures.

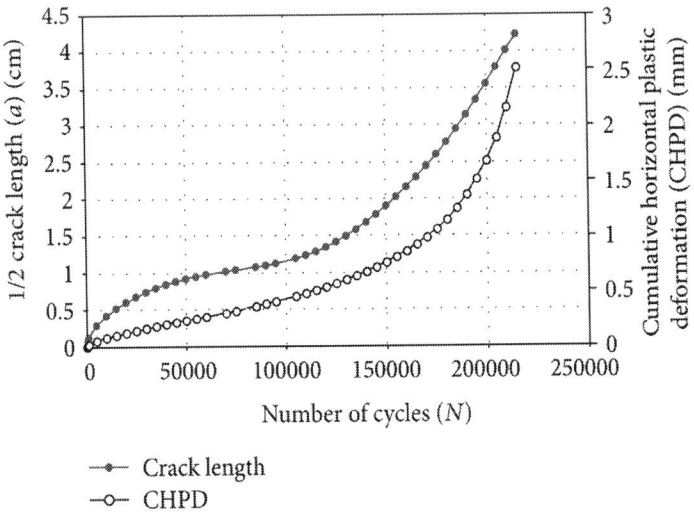

Figure 7: Half-crack length and cumulative HPD as a function of the number of load repetitions for AC10-2% SBR mixture.

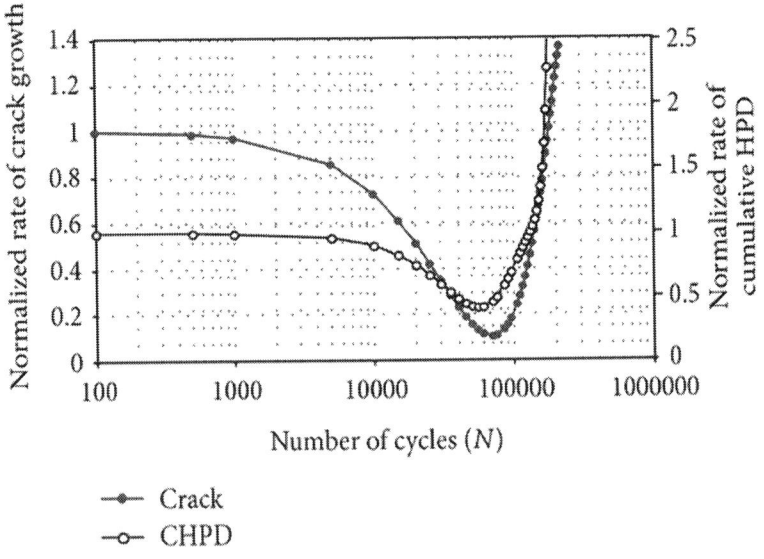

Figure 8: Normalized rate of crack growth da/dN and rate of cumulative HPD (d(CHPD)/ dN) for AC10-2% SBR.

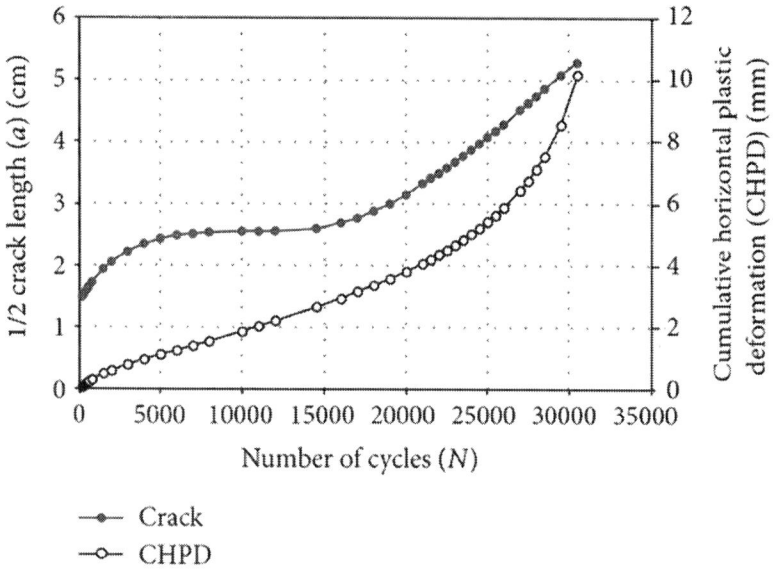

Figure 9: Half-crack length and cumulative HPD as a function of the number of load repetitions for MDOT AC mixture with slot at the center.

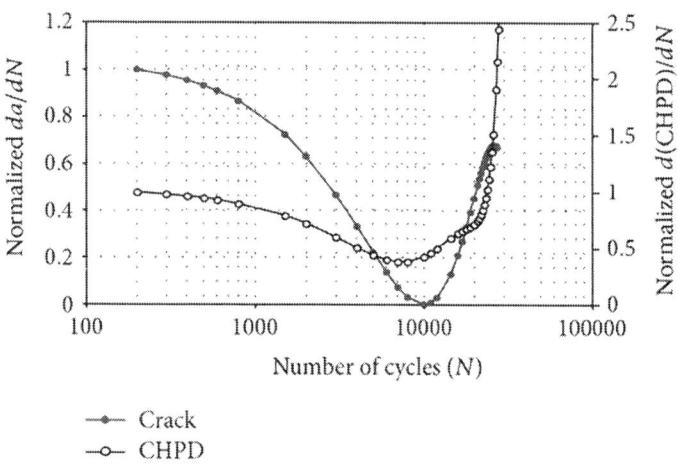

Figure 10: Normalized rate of crack growth da/dN and rate of cumulative HPD ($d(CHPD)/dN$) for MDOT AC mixtures.

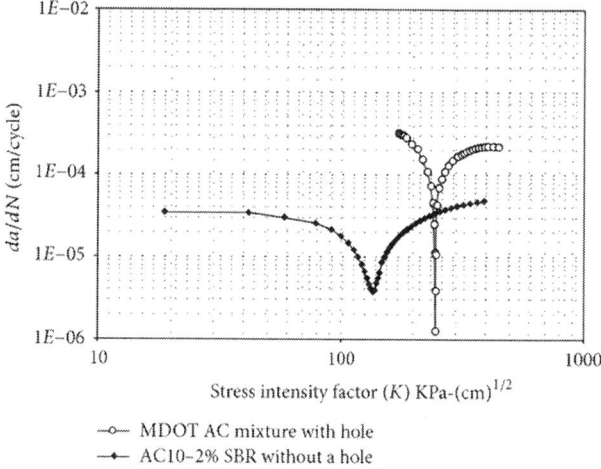

Figure 11: K-da/dN curve for AC10-2% SBR and MDOT AC mixtures with slot at the center.

The decrease in the rate of crack growth may be because there is no single crack, instead, there are many secondary cracks surrounding the main crack. Hence, the crack tip contains many microcracks embedded in the plastic zone. The rate of crack propagation and the path it follows mainly depend on the energy balance at the crack tips. Therefore, the energy due to the applied load is utilized by the main crack as well as the microcracks. Further, if there is a large amount of plastic deformation, the crack tips become blunt. Hence, the presence of microcracks and blunting mechanism at the crack tip reduces the effective stress intensity factor and the rate of crack growth [15, 19].

The test data was analyzed, and the fatigue lives of all samples were obtained using both criteria. Figures 12 and 13 show that the fatigue life obtained from the rate of crack growth criterion was higher than the rate of cumulative HPD criterion. Figure 14 shows the fatigue life calculated using both criteria. It was observed that fatigue life based on the rate of crack growth criterion yielded an average of 45 percent higher values than the rate of accumulation of HPD criterion.

Analysis of Fatigue and Fracture of Hot Mix Asphalt Mixtures

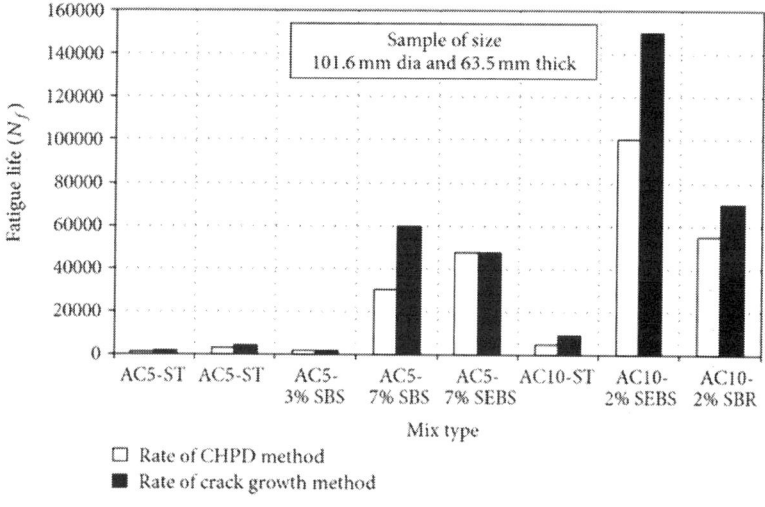

Figure 12: Comparison of the fatigue lives between the rate of cumulative HPD and crack growth criteria for AC straight and PMA mixtures.

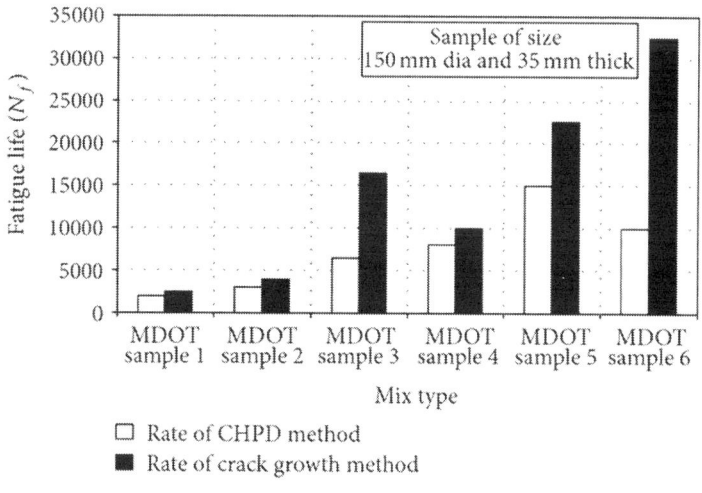

Figure 13: Comparison of the fatigue lives between the rate of cumulative HPD and crack growth criteria for MDOT AC mixture with and without slot at the center.

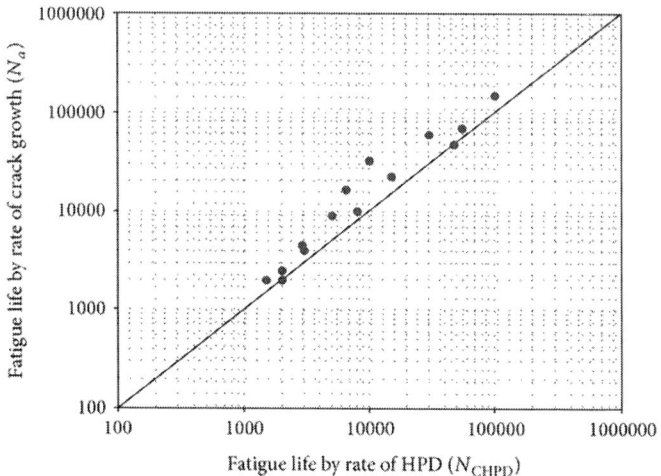

Figure 14: Comparison of the fatigue life between the rate of cumulative HPD and crack growth method.

The difference in the fatigue life obtained from the two criteria can be partly explained by the examination of the crack growth criterion. It is based on theory of elasticity with the assumptions that the material is homogeneous, isotropic elastic solid, and there are negligible changes in the material properties during testing. This is not the case for the HMA mixtures, which are heterogeneous mix of asphalt cement and fine and coarse aggregates. Further, the criterion is based on two-dimensional crack growth. For cylindrical sample tested in indirect cyclic tensile loading, the crack propagates in three dimensions. Given that the crack initiates at the center of the sample, the crack will propagate along both the vertical and horizontal diameters as well as along the thickness of the sample as schematically shown in Figure 15. The crack length equation only calculates the crack length along the vertical diameter of the sample. This may result in higher fatigue life relative to the cumulative HPD criterion. On the other hand the rate of accumulation of HPD criterion of fatigue determination is based on direct measurement of HPD from the test sample. Since the fatigue of cylindrical sample is mainly due to the accumulation of HPD, the rate of accumulation of HPD is a true representation of fatigue crack growth. This criterion produced consistent and repeatable results for both straight and PMA mixtures [5].

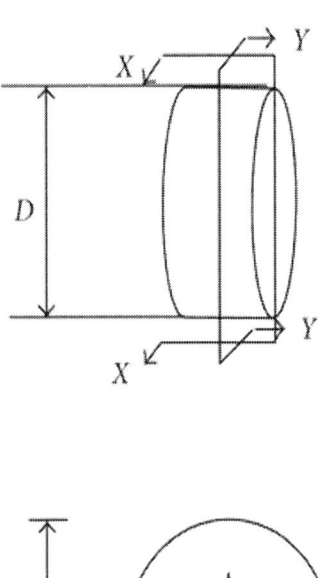

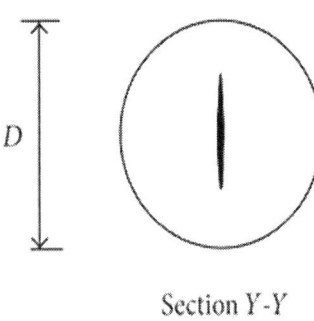

Section Y-Y

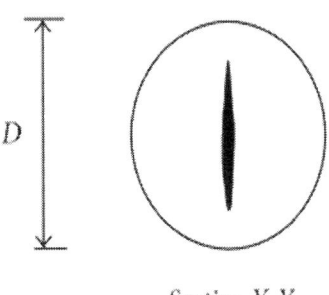

Section Y-Y

Figure 15: Three-dimensional crack propagation model for centrally cracked cylindrical sample under indirect tensile cyclic loading.

CONCLUSIONS

In this paper, a comparison between two different fatigue life criteria for two-dimensional problems represented by cylindrical samples are discussed and presented. One criterion was based on the rate of accumulation of the tensile horizontal plastic deformation (HPD) as a function of the number of load repetitions. The other criterion was based on fracture mechanics and takes into account stress intensity factor and the rate of crack growth with respect to the number of load repetitions. Indirect tensile cyclic loading test was conducted on straight and PMA-modified HMA mixtures with and without slot at the center of sample. It was found that the crack in HMA sample reaches its full length when the compliance is twice of its initial value. The rate of crack growth criterion showed higher fatigue life relative to the rate of accumulation of horizontal plastic deformation. This difference was attributed to the three-dimensional nature of crack propagation. Nevertheless, the rate of accumulation of HPD criterion produced consistent and repeatable results and is recommended to be used for the determination of laboratory fatigue life of HMA mixtures.

ACKNOWLEDGMENTS

The authors wish to express their sincere thanks to the Pavement Research Center of Excellence at Michigan State University and Michigan Department of Transportation for sponsoring this study. Special thanks are also extended to the project review committee for their valuable feedback.

REFERENCES

1. G. Baladi, "Fatigue life and permanent deformation characteristics of asphalt concrete mixes," Transportation Research Record, no. 1227, pp. 75–87, 1989 Y. R. Kim, "Effect of temperature and mixture variables on fatigue life predicted fatigue testing," Transportation Research Record, no. 1317, 1991.

2. A. A. Tayebali, J. B. Sousa, and G. M. Rowe, "Fatigue response of asphalt-aggregate mixtures," Journal of the Association of the Asphalt Paving Technologists, vol. 61, pp. 333–360, 1992.
3. G. Y. Baladi, "Integrated material and structural design method for flexible pavements. Vol. 1," Report. RD-88-109, FHWA, U.S. Department of Transportation, 1988.
4. M. J. Khattak and G. Y. Baladi, "Engineering properties of polymer-modified asphalt mixtures,"Transportation Research Record, no. 1638, pp. 12–22, 1998
5. C. Monismith and J. A. Deacon, "Fatigue of asphalt paving mixtures," Transportation Engineering Journal, vol. 95, no. 2, pp. 317–346, 1969.
6. M. Pell and K. Cooper, "The effect of testing and mix variables on the fatigue performance of bituminous materials," Proceedings of the Association of Asphalt Paving Technologists, vol. 44, pp. 1–37, 1975.
7. J. Alonso, Estudio del proceso de deformación y agrietamiento por fatiga de mezclas bituminosas sometidas a cargas cíclicas [Doctoral Dissertation], Universidad Politécnica de Cataluña, 2006.
8. H. Di Benedetto, C. de la Roche, and L. Francken, "Fatigue of bituminous mixtures: different approaches and RILEM interlaboratory tests," in Proceedings of the 5th International RILEM Symposium on Mechanical Tests for Bituminous Materials (MTBM '97), pp. 15–26, Lyon, France, 1997.
9. H. Baaj, Comportement á la fatigue des matériaux granulaires trités aux liants hydrocarbons [Doctoral Dissertation], INSA, Villeurbanne, France, 2002.
10. O. J. Reyes-Ortiz, A. E. Alvarez-Lugo, and P. Limon, "Effect of the failure criterion on the laboratory fatigue response prediction of hot mix asphalt mixtures," Dyna, vol. 79, no. 174, pp. 31–39, 2012.
11. F. E. Perez, R. Miro, A. Martinez, and J. Alonso, "Desarrollo de un nuevo procedimiento para la evaluación del comportamiento a fatiga de las mezclas bituminosas a partir de su caracterización en un ensayo a tracción," in Primer Premio Internacional a la Innovación en Carreteras-Asociación Española de la Carretera, Madrid, Spain, 2006.

12. J. N. Goodier and N. J. Hoff, "Structural mechanics," in Proceedings of 1st Symposium on Naval Strcutural Mechanics, Pergamon Press, New York, NY, USA, 1960.
13. Y. H. Haung, Pavement Analysis and Design, Prentice Hall, Englewood Cliffs, NJ, USA, 1993.
14. K. Majidzadeh, E. M. Kuaffmann, and D. V. Ramsamooj, "Application of fracture mechanics in the analysis of pavement fatigue," in Proceedings of the Association of Asphalt Paving Technologists, vol. 40, pp. 227–246, 1971.
15. D. Broek, The Practical Use of Fracture Mechanics, Kluwer Academic, Boston, Mass, USA, 1988.
16. D. P. Rooke and D. J. Cartwright, Compendium of Stress Intensity Factor, London Her Majesty's Stationary Office, 1976.
17. J. M. Read and A. Collop, "Practical fatigue characterization of bituminous paving mixtures," inAssociation of Asphalt Paving Technologists Technical Sessions (AAPT ‹97), vol. 66, Salt Lake City, Utah, USA, 1997.
18. R. W. Hertzberg, Deformation and Fracture Mechanics of Engineering Materials, John Wiley & Sons, New York, NY, USA, 1989.

Chapter 2

Dynamic Compressive Deformation and Fracture of a Hollow Bulk Metallic Glass

Haiyan Ye[1,2], Tao Zhang[1,2], Zhihua Wang[3], Hefeng Zhou[1], Shengbo Sang[4], Huijun Yang[1,2] and Junwei Qiao[1,2]

[1]Laboratory of Applied Physics and Mechanics of Advanced Materials, College of Materials Science and Engineering, Taiyuan University of Technology, Taiyuan, China

[2]Key Laboratory of Interface Science and Engineering in Advanced Materials, Ministry of Education, Taiyuan University of Technology, Taiyuan, China

[3]Institute of Applied Mechanics and Biomedical Engineering, Taiyuan University of Technology, Taiyuan, China

[4]MicroNano System Research Center, Taiyuan University of Technology, Taiyuan, China

ABSTRACT

The dynamic mechanical behaviors of hollow $Zr_{41.2}Ti_{13.8}Cu_{12.5}Ni_{10.0}Be_{22.5}$ bulk metal glass (BMG) are investigated using a splitting Hopkinson pressure bar (SHPB) in this study. Upon dynamic compressive loading, the hollow specimen exhibits lower strength and poor ductility, which caused by the higher stress concentration for the hollow one through FEM modeling. The different strain-rate responses for the hollow specimen are compared and explained. On the fracture surface of the hollow samples, there are highly dense vein patterns, many liquid drops and fishbone-like patterns.

INTRODUCTION

Since the successful vitrification of liquid alloys by the rapid solidification in 1960, metallic glasses have received extensive interests [1]. As potential advanced structural materials, bulk metallic glasses (BMGs) have outstanding mechanical properties, such as a high strength of up to 5 GPa, [2] a large elastic deformation limit of around 2 %, as well as good fatigue properties. However, the poor ductility due to the highly-localized inhomogenous deformation and subsequent catastrophic fracture limit their processing and application severely.

$Zr_{41.2}Ti_{13.8}Cu_{12.5}Ni_{10.0}Be_{22.5}$ is one of the best glass formers developed in recent years [3-6]. However, like most other metallic glasses, it also fails catastrophic fracture, due to the excessive propagation of individual shear bands [7,8]. Various methods have been adopted to enhance the toughness of monolithic metallic glasses. For example, a large poisson ratio ( ), which can cause the tip of a shear band in BMGs to extend rather than initial a crack [9]. Deformation is accompanied by the formation of multiple shear bands, which results in an improved plasticity. Besides, it has been demonstrated that an exterior constraint is one of approaches for multiplying the shear bands and preventing the BMGs from the premature fracture [10]. Jiang et al. [11] have investigated the compressive deformation and fracture of hollow BMGs upon quasi-static compressive loading. The results show that the hollow $Zr_{52.5}Cu_{17.9}Ni_{14.6}Al_{10.0}Ti_{5.0}$ (Vit 105) BMG has excellent plasticity by tailoring the stress distribution upon loading. It is obtained that the ductility of BMGs is highly related to different geometries of samples.

In actual engineering applications, most deformation and fracture occur under high-speed dynamic loading, such as defense, aerospace, precision machinery, automotive industries, and high-speed metal forming. However, the resistance to deformation or fracture under dynamic loading is generally lower than that under quasistatic loading, and the plastic deformation is often highly localized in a narrow region [12-16]. Therefore, it is required to obtain information on the dynamic deformation of BMGs so that it can be effectively applied to such strategic fields. Recently, Qiao et al. [17] investigated the dynamic compressive behaviors of Zr-based BMG composites, and found that multiple shear bands were not formed sufficiently under dynamic loading condition, thereby, leading to the lower maximum compressive stress than that measured under quasi-static loading condition. However, the deformation and fracture of hollow Zr-based BMGs under dynamic loading is not yet to be investigated. In this study, the dynamic loading to a hollow BMG will be studied.

EXPERIMENTAL

The $Zr_{41.2}Ti_{13.8}Cu_{12.50}Ni_{10.0}Be_{22.5}$ BMGs in an atomic percent was prepared by arc-melting pure elements with purity higher than 99.9 wt% under highly pure argon atmosphere with titanium as a getter, followed by casting in a water-cooled copper mould, which exhibit high thermal conductivity. The resulting cylindrical BMG rods have a dimension of about 90 mm in length and 6 mm in diameter. The hollow specimens were prepared by drilling machine with an alloy-steel drill bit, and the diameter of drill bit is 2.5 mm. The amorphous nature was confirmed by high-energy X-ray diffraction. The as-cast specimens for dynamic compression tests have a dimension of 6 mm in diameter and 6 mm in height, with an aspect ratio of 1. The ends of the samples were well polished with 1500 grinding paper. The dynamic loading experiments were conducted at room temperature, using a SHPB apparatus, and the detailed process was described elsewhere [18,19]. A SHPB consists of two elastic pressure bars that sandwich the specimen between them, as shown inFigure 1. The striker bar is launched from a gas gun toward the input bar. The impact generates a compressive stress pulse in the incident bar. The stress wave travels toward the specimen, subjecting it to the required stress levels. A portion of the pulse is reflected back into the output bar. The incident and output bars are mounted with strain gages at midway points along the length of the bars to compare the strain signals associated with the waves as they pass by. Upon dynamic compressive loading, specimens were often crushed by the input bar. The fracture surfaces and the lateral surfaces of the deformed samples were investigated to identify the fracture mechanism using scanning electron microscopy (SEM).

RESULTS AND DISCUSSION

Figure 2 shows the synchrotron high-energy X-ray profile of the $Zr_{41.2}Ti_{13.8}Cu_{12.5}Ni_{10.0}Be_{22.5}$ BMGs, together with its corresponding to diffraction pattern (Figure 2 inset). Both indicate a typical amorphous structure for the present Zr-based BMGs, which identifies that no partial crystallization occurred. Thus the effect of crystallization on the mechanical behavior can be excluded.

Figure 3(a) displays the engineering stress-strain curve of the solid $Zr_{41.2}Ti_{13.8}Cu_{12.5}Ni_{10.0}Be_{22.5}$ BMG upon dynamic compressive loading with the strain rate of 2.0×10^3 s^{-1}. It can be seen that the solid Zr-based BMG does not exhibit obvious yielding, which suggests this solid

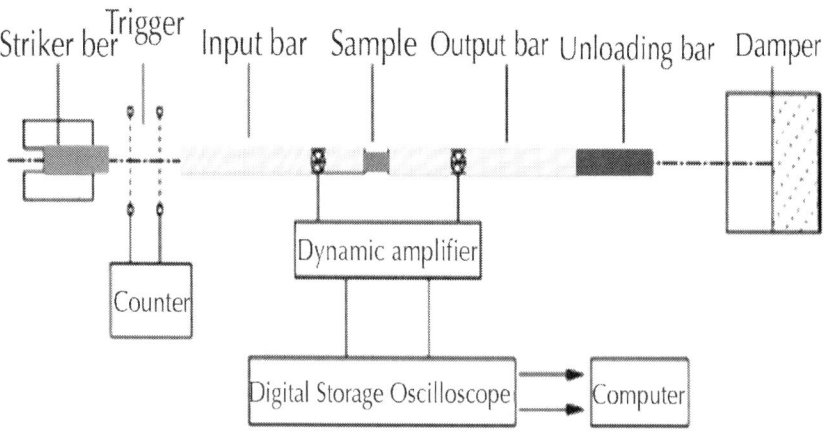

Figure 1: Schematic of the split Hopkinson pressure bar (SHPB).

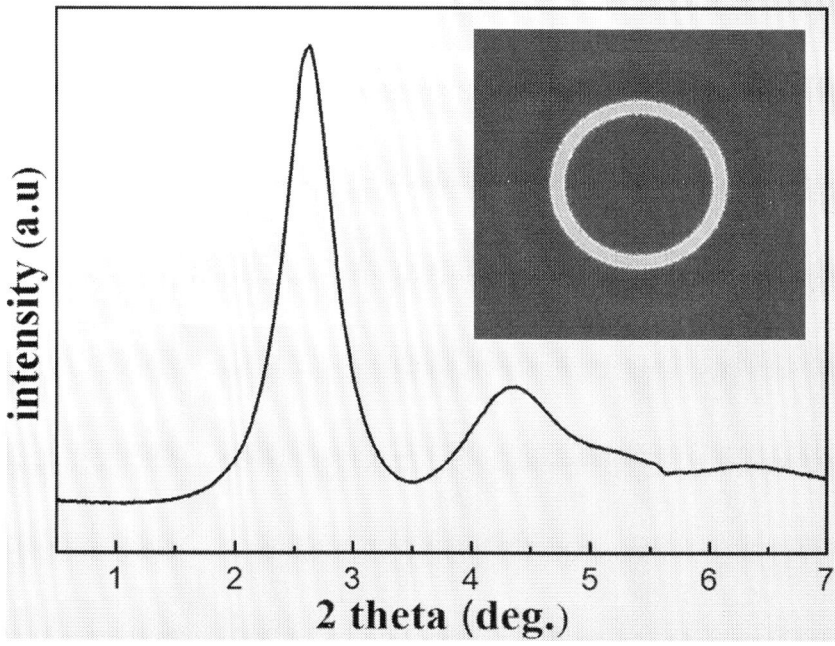

Figure 2: High-energy synchrotron XRD pattern of $Zr_{41.2}Ti_{13.8}Cu_{12.5}Ni_{10.0}Be_{22.5}$ BMGs. Note that the inset is the diffraction pattern of $Zr_{41.2}Ti_{13.8}Cu_{12.5}Ni_{10.0}Be_{22.5}$.

one without obvious plastic deformation, and its fracture strength approaches 1630 MPa. Kim et al. [20] have demonstrated that the solid cylindrical $Zr_{41.2}Ti_{13.8}Cu_{12.5}Ni_{10.0}Be_{22.5}$ BMG's fracture strength is 1440 ± 67 MPa, consistent with the present result. And the fracture strain of solid $Zr_{41.2}Ti_{13.8}Cu_{12.5}Ni_{10.0}Be_{22.5}$ BMG is 1.7%. Jiang et al. have indicated that the fracture strength of solid $Zr_{52.5}Cu_{17.9}Ni_{14.6}Al_{10.0}Ti_{5.0}$ BMG is 1920 MPa upon quasi-static loading, [11] together with poor ductility, which suggests monolithic BMGs have brittle fracture behavior, regardless of under dynamic or quasi-static compressive loading. It should be noted that the solid specimen's elastic deformation stage is not a smooth straight line, indicating that the strain rate is not constant during dynamic compressive process but a mean value.

Figure 3(b) presents the engineering stress-strain curve of the hollow $Zr_{41.2}Ti_{13.8}Cu_{12.5}Ni_{10.0}Be_{22.5}$ BMG upon dynamic compressive loading with the strain rate of 2.0×10^3 s^{-1}. The brittle fracture behaviors are similar to that of the solid one, and its fracture strength is 933 MPa, greatly lower than that of the solid one. The lowing strength for hollow ones has been indicated previously [11]. Compared to the solid one, the hollow specimen exhibits a lower fracture strength. From this curve, it can be seen that the hollow sample has brittle fracture without obvious plasticity upon dynamic loading. But Jiang et al. have reported that the Zr-based hollow specimen exhibits higher plasticity upon quasi-static loading, and the

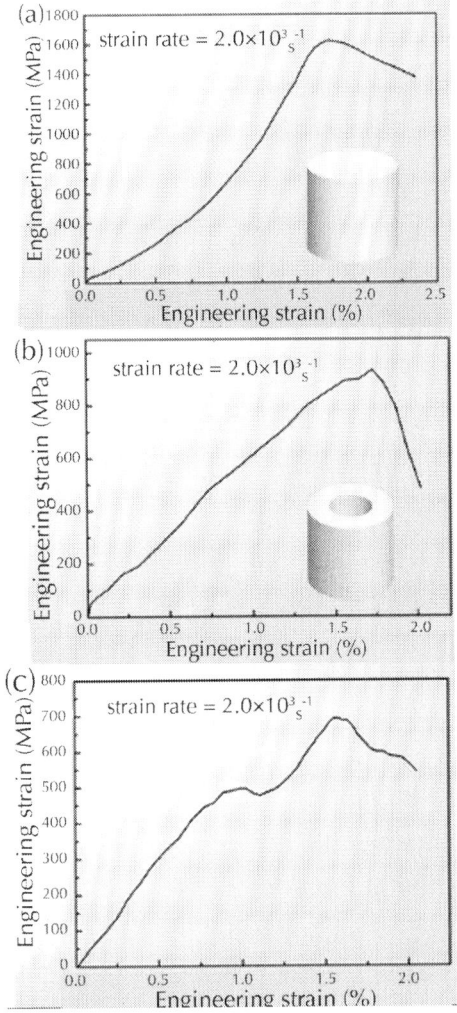

Figure 3: The engineering stress-strain curves of the (a) solid and (b) hollow $Zr_{41.2}Ti_{13.8}Cu_{12.5}Ni_{10.0}Be_{22.5}$ BMGs upon dynamic compressive loading at

the strain rate of $2.0 \times 10^3 s^{-1}$; and (c) hollow $Zr_{41.2}Ti_{13.8}Cu_{12.5}Ni_{10.0}Be_{22.5}$ BMG upon dynamic compressive loading at the strain rate of $1.0 \times 10^3 s^{-1}$. The inset in (a) and (b) are the solid and hollow $Zr_{41.2}Ti_{13.8}Cu_{12.5}Ni_{10.0}Be_{22.5}$ samples.

plastic strain of the hollow portion is as high as 43.7% before the final failure [11]. It is probable that the result is related to combined actions of the applied strain rate, the compression speed, and the propagating speed of the shear bands [21-27]. It follows that BMGs exhibit brittle behavior upon dynamic loading, even if hollow ones exhibit improved plasticity upon quasi-static loading [28].

Figure 3(c) displays engineering stress-strain curve of the hollow $Zr_{41.2}Ti_{13.8}Cu_{12.5}Ni_{10.0}Be_{22.5}$ BMG upon dynamic compressive loading with the strain rate of 1.0×10^3 s^{-1}. It shows an initial elastic deformation, and then a platform is approaching, which indicates a certain plasticity. Finally, another elastic deformation occurs before ultimate fracture, and the slopes of these two elastic deformation portions in the stress-strain curves are approximately equal. Therefore, this shearing process of the hollow specimen can be divided into 3 stages, as shown in Figure 4(a), schematically. Firstly, the wall of specimen (red area) is sheared. Secondly, the hollow potion (yellow area) is sheared. In this stage, the hollow potion could be assumed be made up of a very 'ductile' phaseair. As far as this Zr-based BMG/air composite is concerned, the mechanical behavior is dominated by the deformation of the Zr-based amorphous phase and the special 'ductile' phase. When this 'ductile' phase is sheared, the stress is released instantaneously, and the strain energy is dissipated. In addition, the stress concentration is introduced in the hollow specimen under dynamic loading. In other words, inhomogenous stress distribution contributes to the generation of a certain plasticity. That is why there exists a platform on the stress-strain curve, which is similar to shear hysteresis phenomenon. Thirdly, the side wall of specimen (green area) is sheared. That means the first and third stages are the similar shearing process, which results in the equivalent slope of the two elastic deformation portions.

Compared to the hollow specimen at the stain rate of 1.0×10^3 s^{-1}, the hollow specimen at the stain rate of 2.0×10^3 s^{-1} does not have a shear hysteresis. It seems to undergo one shearing process, as shown in Figure 4(b). To our knowledge, the strength of general ductile alloys are affected by the strain rate hardening, but this stain rate hardening is hardly expected in the hollow sample at the strain rate of 2.0×10^3

s⁻¹. It is explained that even though the plasticity of air is very large, a higher strain rate causes that it does not have sufficient time to occur up to the plastic region. Once initiated, the shear bands propagate, resulting in the prompt occurrence of shear failure. As a result, the plastic stain is not expected. Consequently, the sample exhibits brittle fracture. Different deformation and fracture mechanisms dominate the failure of the hollow specimens at different stain rate, resulting in the different mechanical performances.

At quasi-static loading, the geometrical constraint in the hollow BMG specimen can lead to the multiplication of shear bands, which moderates the localization of plastic deformation and delays the fracture. With the increase of strain rates, the increase of stress concentration is more and more severe. Figure 5 shows the finite-element modeling (FEM) to examine the stress distribution along the shearing plane with a 43 degree with respect to the loading direction. The external forcing is the same for the solid in Figure 5(a) and hollow one in Figure 5(b). It can be seen that the maximum stress concentrates at the center of the shearing plane for the solid Zr-based BMG. In contrast, for the hollow one, the high stress gradient is in the vicinity of the hole, and the stress concentration.

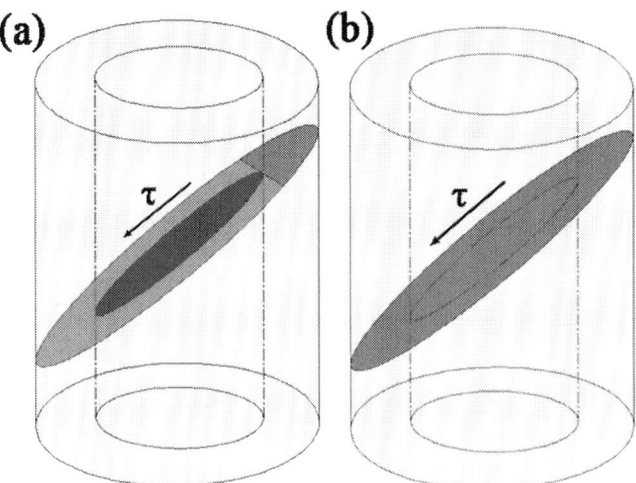

Figure 4: The schematic shearing processes of the hollow $Zr_{41.2}Ti_{13.8}Cu_{12.5}Ni_{10.0}Be_{22.5}$ specimens at the strain rate of (a) 1.0×10^3 s⁻¹ and at the strain rate of (b) 2.0×10^3 s⁻¹.

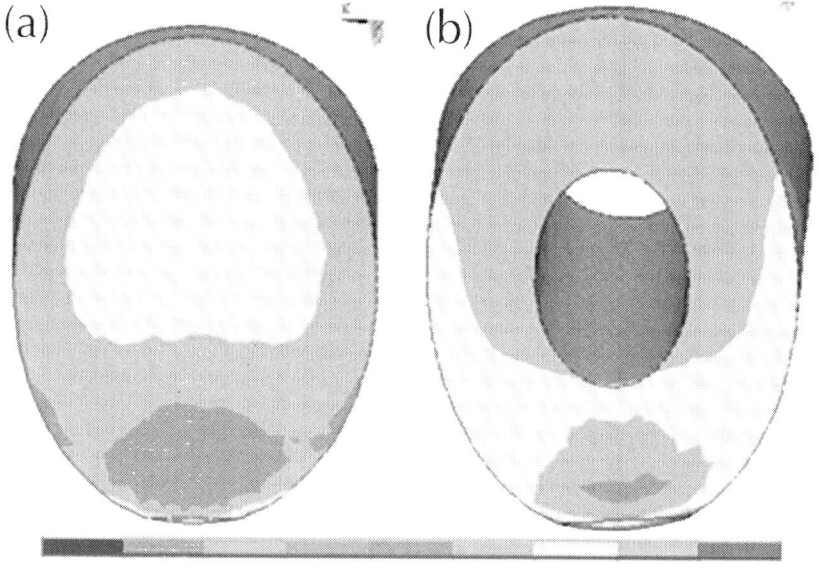

Figure 5: The finite-element modeling (FEM) to examine the stress distribution along the shearing plane.

occurs near the hole. At the same external loading, the stress in the shearing plane of hollow ones is greatly higher than that of solid ones. As a result, an early failure happens with lower strengths for the hollow ones, in agreement with the experimental re sults.

Figure 6(a) shows the SEM image of the fracture surface of solid deformed specimen upon dynamic loading. The sample cracks into pieces after instantaneous destruction, as shown in the inset of Figure 6(a), caused by re-loading to the failure samples. It can be seen that typical vein patterns are distributed on the fracture surface, which has been widely observed in the fracture surfaces of other BMGs [29]. This phenomenon is very different from that under quasi-static loading, and the quasi-static fractured surface is almost covered by many liquid droplets and few vein patterns [30]. Upon quasistatic, the energy accumulated has sufficient time to be converted into heat, resulting in the temperature rise. So the vein patterns prevail on the dynamically fractured surface due to the insufficient temperature rise. When the bright dotted region in Figure 6(a) is magnified, a number of vein patterns are still observed, but tend to become more elongated, as shown in Figure 6(b). In addition, there are some resolidified liquid droplets (marked by the dark arrows) can be observed, which indicates that adiabatic heating occurs before yielding in this $Zr_{41.2}Ti_{13.8}Cu_{12.5}Ni_{10.0}Be_{22.5}$ BMG. However, Bruck et al.

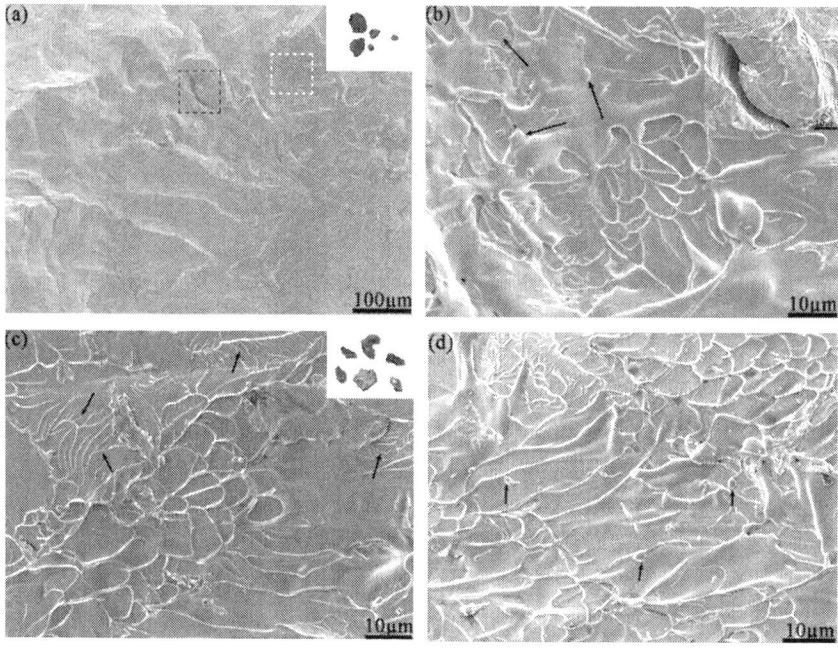

Figure 6: (a) Fracture surface of the solid BMG specimen and (b) enlarged images of the bright and dark dotted region in (a); (c) and (d) Fracture surface of the hollow BMG specimen. The inset in (a) and (c) are the images of solid and hollow fractured samples respectively.

observed the large temperature rise occurred before yielding in a Zr-based BMG upon dynamic compressive loading [31]. The difference may be related with different strain rates and chemical compositions [32,33]. The present experiment features the resolidified droplets, which suggests the temperature rise may be higher than the melting temperature at the moment of fracture upon dynamic loading. As a result, Zr-based amorphous phase only occurs localized melting with a relatively high strain rate, and the resolidified droplets can not cover the entire dynamically fractured surface, thus the vein patterns are still observed. When the dark dotted region in Figure 6(a) is magnified, besides vein patterns and resolidified droplets, a microcrack can be observed due to the severe deformation, as seen in the inset of Figure 6(b), consistent with recent theoretical and experimental studies [34,35]. It has demonstrated that the temperature rise in the narrow plastic zone, associated with a moving crack within the shear bands, can be as high as a few thousand Kelvin. This phenomenon can be explained with the energy conversion. Kinetic energy changes into heat and surface energy during this dynamic impact process. Just as some large cracks, they absorb much surface energy.

Figures 6(c) and (d) display the SEM micrographs of the broken hollow specimen surface upon dynamic loading. The sample cracks into pieces after instantaneous failure, as shown in the inset ofFigure 6(c). Except for the elongated vein-like patterns, many fishbone-like patterns are observed on the fracture surface, marked by the dark arrows in Figure 6(c), and analogous result has been found in other BMGs [36]. It demonstrated that the fishbone-like patterns as well as the smooth regions between them indicate a lower crack propagation speed and a smaller temperature rise.Figure 6(d) shows some solidified liquid droplets (labeled by the dark arrows), which are distributed on the edges of veins, suggesting a significant temperature rise.

CONCLUSIONS

After the text edit has been completed, the paper is ready The dynamic deformation behaviors of $Zr_{41.25}Ti_{13.75}Cu_{12.50}Ni_{10.00}Be_{22.50}$ BMGs are investigated. The hollow and solid specimens' deformation and fracture behaviors upon dynamic compressive loading are compared. They both exhibit little ductility, which are similar to other monolithic BMGs. Their fracture strengths are 1630 MPa and 933 MPa, respectively. It is obvious that the hollow one's fracture strength is lower than that of the solid one, which is caused by the higher stress concentration for the hollow one through FEM modeling. The hollow specimens' dynamic mechanical behaviors are very different from that upon quasi-static loading. The former exhibits poor plasticity, but the later exhibits large plasticity. The different strain-rate responses for the hollow Zr-based BMGs samples are compared and deeply explained. The vein patterns on the dynamically fractured surface are incompletely covered by the liquid drops, due to the insufficient temperature rise.

ACKNOWLEDGEMENTS

J.W.Q. and S.B.S. would like to acknowledge the financial support of National Natural Science Foundation of China (No.51101110, 51371122, and 51105267), the Youth Science Foundation of Shanxi Province, China (No.2012021018-1 and 2012021013-1), Research Project Supported by Shanxi Scholarship Council of China (No.2012-032 and 2012-030), and Supported by Program for the Outstanding Innovative Teams of Higher Learning Institutions of Shanxi (2013). Z.H.W. would like to acknowledge the Program for New Century Excellent Talents in University (NCET-10-0930).

REFERENCES

1. W. Klement, R. H. Willens and P. Duwez, "Non-crystalline Structure in Solidified Gold-Silicon Alloys," Nature, Vol. 187, No. 4740, 1960, pp. 869-870. doi:10.1038/187869b0.
2. A. Inoue, B. L. Shen, H. Koshiba, H. Kato and A. R. Yavari, "Cobalt-Based Bulk Glassy Alloy with Ultrahigh Strength and Soft Magnetic Properties," Nature Materials, Vol. 2, No. 10, 2003, pp. 661-664. doi:10.1038/nmat982.
3. A. Peker and W. L. Johnson, "A Highly Processable Metallic Glass," Applied Physics Letters, Vol. 63, No. 17, 1993, pp. 2342-2344. doi:10.1063/1.110520.
4. A. Inoue, T. Zhang, N. Nishiyama, K. Ohba and T. Masumoto, "Preparation of 16 mm Diameter Rod of Amorphous $Zr_{65}Al_{7.5}Ni_{10}Cu_{17.5}$ Alloy," Materials Transactions, Vol. 34, No. 12, 1993, pp. 1234-1237.
5. W. H. Wang, C. Dong and C. H. Shek, "Bulk Metallic Glasses," Materials Science and Engineering: R, Vol. 44, No. 2-3, 2004, pp. 45-89. doi:10.1016/j.mser.2004.03.001.
6. H. G. Kim and H. J. Jang, " Effect of Temperature and PH Level on the Corrosion Behavior of Amorphous $Co_{69}Fe_{4.5}Nb_{1.5}Si_{10}B_{15}$ Alloy," Metals and Materials International, Vol. 16, No. 4, 2010, pp. 581-585. doi:10.1007/s12540-010-0810-4
7. H. Kimura and T. Masumoto, "Strength, Ductility and Toughness—A Model Study in Mechanics," In: F. E. Luborsky,

Ed., Amorphous Metallic Alloys, Butterworths, London, 1983, p. 187.

8. F. Spaepen and A. I. Taub, "Flow and Fracture," In: F. E. Luborsky, Ed., Amorphous Metallic Alloys, Butterworths, London, 1983, p. 231.

9. Z. F. Zhang, H. Zhang, X. F. Pan, J. Das and J. Eckert, "Effect of Aspect Ratio on the Compressive Deformation and Fracture Behaviour of Zr-Based Bulk Metallic Glass," Philosophical Magazine Letters, Vol. 85, No. 10, 2005, pp. 513-521. doi:10.1080/09500830500395237.

10. J. Lu and G. Ravichandran, "Pressure-Dependent Flow Behavior of $Zr_{41.2}Ti_{13.8}Cu_{12.5}Ni_{10}Be_{22.5}$ Bulk Metallic Glass," Journal of Materials Research, Vol. 18, No. 9, 2003, pp. 2039-2042.

11. W. H. Jiang, K. Qiu, F. Liu, H. Choo and P. K. Liaw, "Compressive Deformation and Fracture of a Hollow Bulk-Metallic Glass," Advanced Engineering Materials, Vol. 9, No. 3, 2007, pp.147-150. doi:10.1002/adem.200600249.

12. D.-G. Lee, S. Lee and C. S. Lee, "Quasi-Static and Dynamic Deformation Behavior of Ti-6Al-4V Alloy Containing Fine 2-Ti3Al Precipitates," Materials Science and Engineering: A, Vol. 366, No. 1-2, 2004, pp. 25-37. doi:10.1016/j.msea.2003.08.061.

13. A. Marchand, J. Duffy and J. Mech, "An Experimental Study of the Formation Process of Adiabatic Shear Bands in a Structural Steel," Journal of the Mechanics and Physics of Solids, Vol. 36, No. 3, 1988, pp. 251-283. doi:10.1016/0022-5096(88)90012-9

14. K. A. Hartley, J. Duffy and R. H. Hawley, "Measurement of the Temperature Profile during Shear Band Formation in Steels Deforming at High Strain Rates," Journal of the Mechanics and Physics of Solids, Vol. 35, No. 3, 1987, pp. 283-301. doi:10.1016/0022-5096(87)90009-3.

15. D.-K. Kim, S. Y. Kang, S. Lee and K. J. Lee, "Analysis and Prevention of Cracking Phenomenon Occurring during Cold Forging of Two AISI 1010 Steel Pulleys," Metallurgical and Materials Transactions A, Vol. 30, No. 1, 1999, pp. 81-92. doi:10.1007/s11661-999-0197-3.

16. K. Cho, S. Lee, S. R. Nutt and J. Duffy, "Adiabatic Shear Band Formation during Dynamic Torsional Deformation of an HY-100

Steel," Acta Metallurgica, Vol. 41, No. 3, 1993, pp. 923-932. doi:10.1016/0956-7151(93)90026-O

17. J. W. Qiao, P. Feng, Y. Zhang, Q. M. Zhang and G. L. Chen, "Quasi-Static and Dynamic Deformation Behaviors of Zr-Based Bulk Metallic Glass Composites Fabricated by the Bridgman Solidification," Journal of Alloys and Compounds, Vol. 486, No. 1-2, 2009, pp. 527-531. doi:10.1016/j.jallcom.2009.06.196.

18. D. G. Lee, Y. G. Kim, B. Hwang, S. Lee and Y. T. Lee, "Effects of Temperature on Dynamic Compressive Properties of Zr-Based Amorphous Alloy and Composite," Materials Science and Engineering: A, Vol. 472, No. 1-2, 2008, pp. 316-323. doi:10.1016/j.msea.2007.04.050.

19. Y. F. Xue, H. N. Cai, L.Wang, F. C. Wang, H. F. Zhang and Z. Q. Hu, "Deformation and Failure Behavior of a Hydrostatically Extruded $Zr_{38}Ti_{17}Cu_{10.5}Co_{12}Be_{22.5}$ Bulk Metallic Glass/Porous Tungsten Phase Composite under Dynamic Compression," Composites Science and Technology, Vol. 68, No. 15-16, 2008, pp. 3396-3400.doi:10.1016/j.compscitech.2008.09.026.

20. Y. G. Kim, S. Y. Shin, J. S. Kim, H. Huh, K. J. Kim and S. Lee, "Dynamic Deformation Behavior of Zr-Based Amorphous Alloy Matrix Composites Reinforced with STS304 or Tantalum Fibers," The Minerals, Metals & Materials Society and ASM International, Vol. 43, No. 9, 2012, pp 3023-3033.

21. Y. F. Xue, L. Wang, X. W. Cheng, F. C. Wang , H. W. Cheng, H. F. Zhang and A. M. Wang, "Strain Rate Dependent Plastic Mutation in a Bulk Metallic Glass under Compression," Materials & Design, Vol. 36, No. 4, 2012, pp. 284-288.doi:10.1016/j.matdes.2011.11.025.

22. F. Spaepen, "A Microscopic Mechanism for Steady State Inhomogeneous Flow in Metallic Glasses," Acta Metallurgica, Vol. 25, No. 4, 1977, pp. 407-415. doi:10.1016/0001-6160(77)90232-2.

23. A. S. Argon, "Plastic Deformation in Metallic Glasses," Acta Metallurgica, Vol. 27, No. 1, 1979, pp. 47-58. doi:10.1016/0001-6160(79)90055-5

24. W. H. Jiang and M. Atzmon, "Rate Dependence of Serrated Flow in a Metallic Glass," Journal of Materials Research, Vol. 18, No. 4, 2003, pp. 755-757.doi:10.1557/JMR.2003.0103.

25. C. A. Schuh, A. S. Argon, T. G. Nieh and J. Wadsworth, "The Transition from Localized to Homogeneous Plasticity during Nanoindentation of an Amorphous Metal," Philosophical Magazine, Vol. 83, No. 22, 2003, pp. 2585- 2597. doi:10.1080/14786430310001 18012.
26. C. A. Schuh, A. C. Lund and T. G. Nieh, "New Regime of Homogeneous Flow in the Deformation Map of Metallic Glasses: Elevated Temperature Nanoindentation Experiments and Mechanistic Modeling," Acta Materialia, Vol. 52, No. 20, 2004, pp. 5879-5891.doi:10.1016/j.actamat.2004.09.005.
27. G. P. Zhang, W. Wang, B. Zhang, J. Tan, C. S. Liu, "On rate-Dependent Serrated Flow Behavior in Amorphous Metals during Nanoindentation," Scripta Materialia, Vol. 52, No. 11, 2005, pp. 1147-1151. doi:10.1016/j.scriptamat.2005.01.045
28. J. W. Qiao, P. Feng, Y. Zhang, Q. M. Zhang, P. K. Liaw, G. L. Chen, "Quasi-Static and Dynamic Deformation Behaviors of in Situ Zr-Based Bulk-Metallic-Glass-Matrix Composites," Journal of Materials Research, Vol. 25, No. 12, 2010, pp. 2264-2270. doi:10.1557/jmr.2010.0289.
29. G. Wang, D. Q. Zhao, H. Y. Bai, M. X. Pan, A. L. Xia, B. S. Han, X. K. Xi, Y. Wu and W. H. Wang, "Nanoscale Periodic Morphologies on the Fracture Surface of Brittle Metallic Glasses," Physical Review Letters, Vol. 98, No. 23, 2007, Article ID: 235501. doi:10.1103/PhysRevLett.98.235501.
30. C. T. Liu, L. Healtherly, D. S. Easton, C. A. Carmichael, J. H. Schneibel, C. H. Chen, J. L. Wright, M. H. Yoo, J. A. Horton and A. Inoue, "Test Environments and Mechanical Properties of Zr-Base Bulk Amorphous Alloys," Metallurgical and Materials Transactions A, Vol. 29, No. 7, 1998, pp. 1811-1820. doi:10.1007/s11661-998-0004-6.
31. H. A. Bruck, A. J. Rosakis and W. L. Johnson, "The Dynamic Compressive Behavior of Beryllium Bearing Bulk Metallic Glasses," Journal of Materials Research, Vol. 11, No. 2, 1995, p. 503. doi:10.1557/JMR.1996.0060
32. G. Subhash, R. J. Dowding and L. J. Kecskes, "Characterization of Uniaxial Compressive Response of Bulk Amorphous Zr-Ti-Cu-Ni-Be Alloy," Materials Science and Engineering: A, Vol. 334, No. 1-2, 2002, pp. 33-40. doi:10.1016/S0921-5093(01)01768-3.

33. G. Sunny, F. Yuan, V. Prakash and J. Lewandowski, "Effect of High Strain Rates on Peak Stress in a Zr-Based Bulk Metallic Glass," Journal of Applied Physics, Vol. 104, No. 9, 2008, Article ID: 093522. doi:10.1063/1.3009962
34. J. J. Lewandowski and A. L. Greer, "Temperature Rise at Shear Bands in Metallic Glasses," Nature Materials, Vol. 5, No. 1, 2005, pp. 15-18.
35. B. Yang, M. L. Morrison, P. K. Liaw, R. A. Buchanan, G. Y. Wang, C. T. Liu and M. Denta, "The Electrochemical Evaluation of a Zr-Based Bulk Metallic Glass in a Phosphate-Buffered Saline Electrolyte," Applied Physics Letters, Vol. 86, No. 3, 2005, Article ID: 141904.
36. B. A. Sun, J. Tan, S. Pauly, U. Kuhn and J. Eckert, "Stable Fracture of a Malleable Zr-Based Bulk Metallic Glass," Journal of Applied Physics, Vol. 112, No. 10, 2012, Article ID: 103533. doi:10.1063/1.4767327.

Chapter 3

High Sensitive Methods for Health Monitoring of Compressor Blades and Fatigue Detection

Mirosław Witoś

Air Force Institute of Technology (AFIT), Ks. Bolesława 6, 01-494 Warszawa, Poland

ABSTRACT

The diagnostic and research aspects of compressor blade fatigue detection have been elaborated in the paper. The real maintenance and overhaul problems and characteristic of different modes of metal blade fatigue (LCF, HCF, and VHCF) have been presented. The polycrystalline defects and impurities influencing the fatigue, along with their related surface finish techniques, are taken into account. The three experimental methods of structural health assessment are considered. The metal magnetic memory (MMM), experimental modal analysis (EMA) and tip timing (TTM) methods provide information on the damage of diagnosed objects, for example, compressor blades. Early damage symptoms, that is, magnetic and modal properties of material strengthening and weakening phases (change of local dislocation density and grain diameter, increase of structural and magnetic anisotropy), have been described. It has been proven that the shape of resonance characteristic gives abilities to determine if fatigue or a blade crack is concerned. The capabilities of the methods for steel and titanium alloy blades have been illustrated in examples from active and passive experiments. In the conclusion, the MMM, EMA, and TTM have been verified, and the potential for reliable diagnosis of the compressor blades using this method has been confirmed.

INTRODUCTION

Many different fatigue failures,
- low cycle fatigue (LCF)
- high cycle fatigue (HCF),
- very high cycle fatigue (VHCF),
- thermomechanical fatigue (TMF),

could occur throughout the turbine engine's life (Figure 1). Most of them are damage to compressor blades described in the paper.

Figure 1: Problems of structural fatigue in turbine engines.

Fatigue cracks propagating in rotor blades, the incorrect control of the engine's fuel system, and the lack of knowledge on the loads affecting the bearing system generally could cause formidable hazard to flight safety (Figure 2), as well as to engine life and reliability. Therefore, the AFIT keeps looking for new methods to recognize engineering, manufacturing, overhaul and service errors as well as stochastic overloads during the engine's running. Recognition of operational and maintenance problems is the first step to actively diagnose and control fatigue progress (engine's structural durability and reliability) as well as for prolongation engine time between overhaul and TBO [1–3].

Varig Boeing 767-241ER PP-VNN at Sao Paulo-Guarulhos 07 JUN 2000/ASN

Figure 2: CF6-80 engine: fatigue destruction of III–IX stages of HP compressor [24].

The paper presents three diagnostic methods.
- A metal magnetic memory [4–7], which has been used as a sensitive passive observer of residual and applied stresses and material damages (of ferromagnetic and some austenite parts) [8–15]. Nowadays in the world MMM method has been used only for NDT of industrial objects, for example, welding joints,

gas and steam turbine parts, rope, underground pipes, supporting structures in order to increase their TBO. The method has been tested in Poland for aviation (as NDE and SHM applications) since 2008 [3,15, 16].

- An experimental modal analysis [17–21], which has been used in Poland as a sensitive NDE method during overhaul blade tests since 2008 and the method of the high-frequency identification of blade modal properties [3, 22, 23].
- A tip timing method [25–33], which is one of the most interesting methods of complex diagnosing of jet engines and a powerful tool in order to investigate dynamic phenomena. The method has been used in the Polish Air Force since 1993 with the SNDŁ-1b/SPŁ-2b diagnosing system developed for the SO-3 jet engines. Since 1997, this method has been also used in Poland in the postrepair/postoverhaul acceptance tests. Now, the method is developed as digital board computer for structural health monitoring (SHM) in different aero-engines. Roots of the method date back to years 20 of the 20th century, when it was drawn up in the analogue version by Sir Campbell to needs of examining vibration of steam turbine blades [34]. At present the method is dynamically elaborated and used in the world, mainly in aviation, in the digital version. Standardizing works (ISA 107.1 subcommittee) are being conducted by partner of European Virtual Institute for Gas Turbine Instrumentation (EVI GTI) [35] and the Propulsion Instrumentation Working Group (PIWG) [36]. Organizations joining the largest producers of aero-engines, research and education units and companies of the metrological support. Details of the TTM application are being protected by numerous inventions, for example, [37–54].

The above described methods are sources of complex information about blade quality (of design, production, and overhaul) and real dynamics of phenomena correlated by modal properties, which have an effect on blades damage and fatigue differentiation. This information is used for holistic analysis of fatigue problems in aeroengines with influence of human and operating factors and actively control fatigue of compressor blade (Figure 3) as well as for verification of a FEM model.

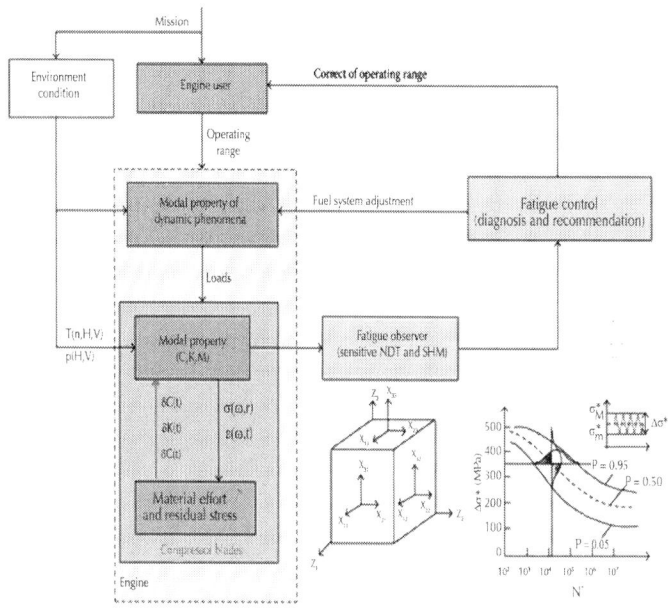

Figure 3: Idea of actively control fatigue by engine user—do not wait for blade crack [3].

MOTIVATION

In the years 1975–91 as many as 25 first-stage compressor blades of ten SO-3 jet engines suffered fatigue-attributable break-offs, which caused two accidents. The metallographic examination of damaged blades made out of the 18H2N4WA alloy structural steel has proved that the crack initiation zone was located either on the leading edges (55%) or on the blade-back surfaces (45%), in the areas of nodal lines of the first mode vibration. Crack propagation occurred at low-level stresses (HCF problem) or high-level stresses (LCF problem) (Figure4). Fatigue fracture covering as much as 95% of the blade's cross section was found in one of the blades. Furthermore, it has also been found that erosion and corrosion, both occurring on the blade's face surface, as well as fine mechanical damages on the leading edge are stress concentrators [55]. Fatigue problem was also observed in titanium blades (Ti5.8Al-3.7Mo) in the TW3-117 engines in the years 2005–2007 [3].

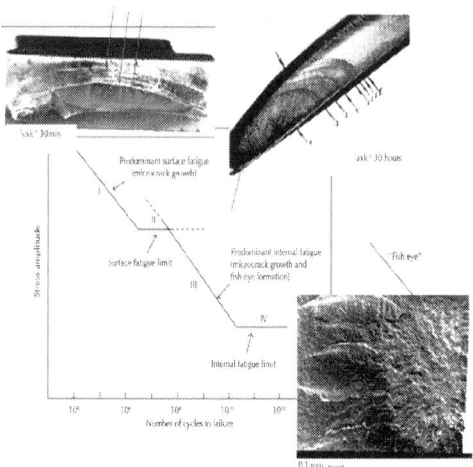

Figure 4: Fatigue problems (LCF, HCF, and VHCF) of compressor blades [3, 56].

The gigacycle fatigue of compressor blade (VHCF problem) with "fish eye" symptoms under the blade surface has been observed at foreign users, for example, in Russia [56, 57] (Figure 4). Compressor blades run a risk of VHCF problem because they count more than 3.10^9 cycles for 1st flexible mode and more than 1.10^{10} cycles for 1st torsional mode during TBO. High risk of VHCF problem, with crack nucleation under the blade surface and stresses level below surface fatigue limits, concerns of high resistant material and blades made with surface finish techniques [58–60].

Uncontrolled blade over fatigue
- is a threat for service safety;
- limits aeroengine life time;
- increases maintenance costs.

It is also a great challenge for diagnostics engineers. Specialists, who are familiar with abilities of the NDE and SHM methods and continuum damage mechanics of material (Figure 5), are searching for answers to questions of the aero-engine user and the technologist of the renovation plant.

- Level I—before the first compressor blade crack or break-off

Which types of aero-engine are unrecognized compressor fatigue problem on?

⇓

Is it possible to assess the fatigue risk of blades through the casing of the compressor?
- Level II—technology and the quality of overhaul

What methods of the compressor blades verification should apply in the overhaul so that the blade does not break in the guarantee period?

⇓

How to recognize overload and weakened compressor blades (2nd stage of damage)?
- Level III—preventive activity in the overhaul and maintenance

If the crack of the blade was detected, what is causing for fatigue problem?

⇓

What mistakes of the engine use and repair are affecting for precipitated fatigue of compressor blades?

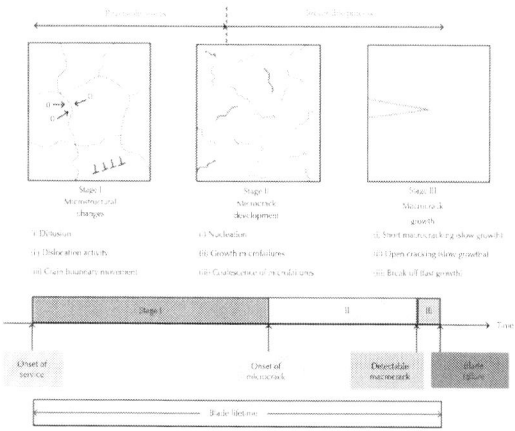

Figure 5: Continuum damage mechanics for compressor blade material. The first stage and the beginning of the second stage of damage are copying reversible changes [3].

Classical NDE methods (X-ray, eddy current, ultrasound, magnetic particle and fluorescent) are very low effective ones for diagnosing blade crack before 3rd stage of damage because of the following.

- They do not answer the questions mentioned above. Finding the crack or other discontinuity of the structure, rather than its causes, is their main task.
- Crack gap closing during engine standstill: Theoretical possibilities of the NDE method (detection of defects about min. size 0.3 mm) turn out to be false during examinations of the blades carried out after the longer stop of the engine, (Figure 6).
- Lack of reliable information about real operating conditions.
- Lack of knowledge about early cracking symptoms and mechanisms.
- Limited access to tested blades (because of inlet stator vane).

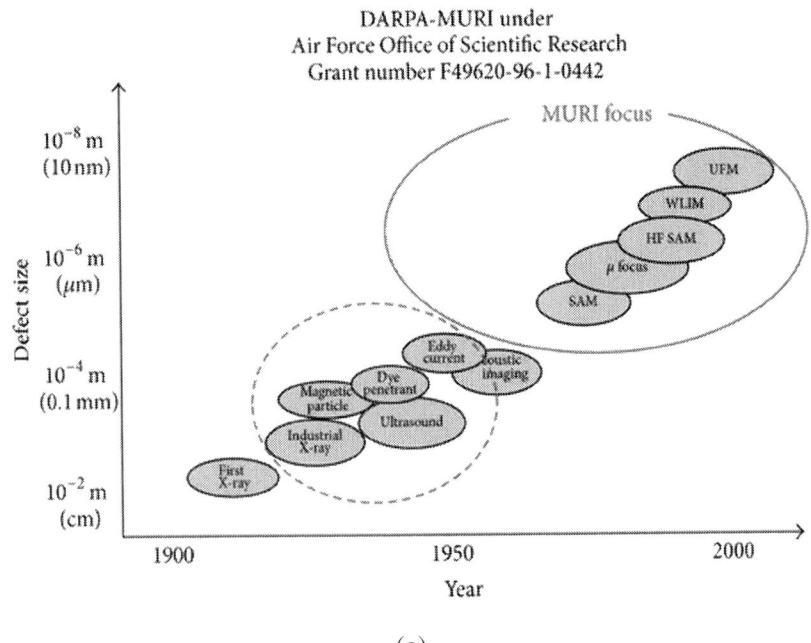

(a)

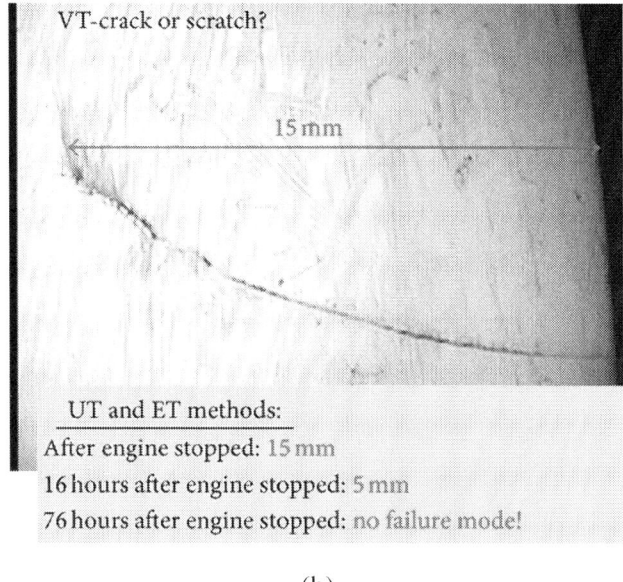

(b)

Figure 6: Showing: (a) Contemporary research capabilities of material degradation in service (the classic NDT methods) and the laboratory (MURI focus) [55]; (b) Influence residual stresses on closing the crack gap and the change of reading by ultrasound and eddy current methods after switching the engine off.

Other disadvantage of the NDE methods in use (during overhaul and service) is no possibility of fatigue prognosis.

This disadvantage is very important for blades having fatigue problems, for example, the 1st compressor blades of the SO-3 engine. These blades have design errors—too low the first-mode mistuning from the 2nd rotational harmonic excitation. Therefore, too high stress and fast fatigue crack initiation can occur during operation. These conditions take place during the take-off phase when there is a foreign object (e.g., bird) lying in the inlet or the inlet icing occurs (Figure 7). Under such conditions time between crack initiation and blade damage can be shorter than time of a single flight (LCF problem). Disturbance of the pressure and velocity in the intake are being moved by the entire length of the compressor. Hazardous blade vibrations occur only at synchronous resonance if the level of excitations is greater than acceptable. A diagram of Campbell is outlining zones of the synchronization, but it is well known only to a designer of the compressor. Flutter, surge (source of asynchronous resonances), and foreign object damage (FOD, local stress concentrator) are also source of LCF problems. Surge and FOD are easily detected by the aeroengine users.

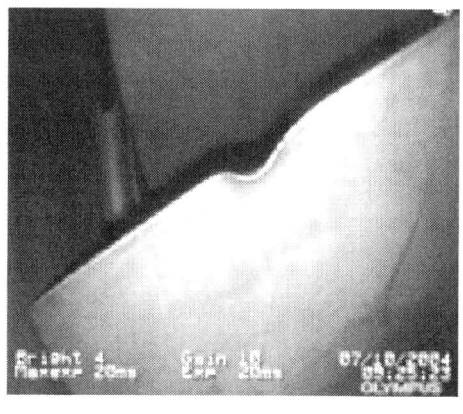

(a)

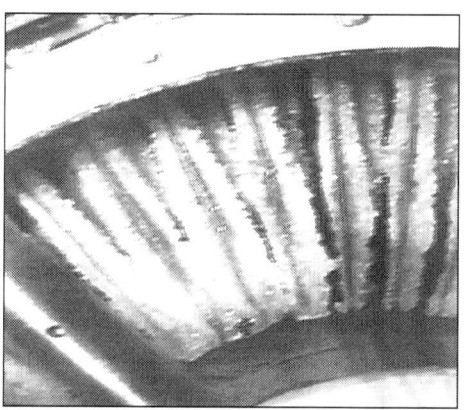

(b)

Figure 7: Showing LCF concentrations: (a) blade with FOD; (b) icing of compressor inlet.

Some errors of setting or hidden defects of the fuel control system affect the long-term working of compressor near the unstable limits (Figure 8) (stall, flutter, or surge during acceleration and deceleration, disturbance of the temperature field in compressor, combustion chamber and turbine), as well as high level of rotor unbalancing and alignment are the source of generating hidden fatigue problems of compressor blades (HCF and VHCF problems). Synchronous and asynchronous resonances are appearing in these working conditions of the compressor. The lack of distinct manifestations of the engine during disorders mentioned above causes that HCF and VHCF problems of blades, including tearing them off, are surprising for the user. Initiating the opened crack (3rd stage of damage) and its propagation are taking place at the low vibration amplitude of blades. Decreasing risk level of blades HCF and LCF is possible by correct overhaul and operation errors, that is, shape of operating area changes [2, 3, 61–63].

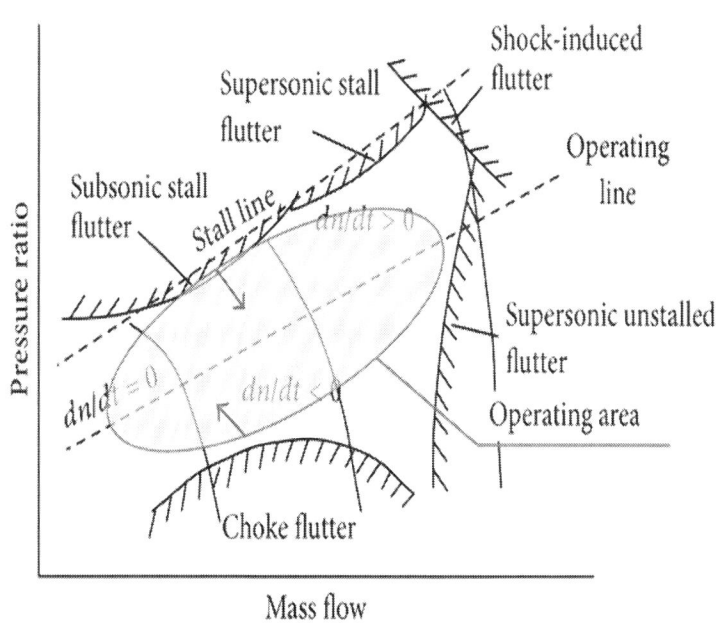

(a)

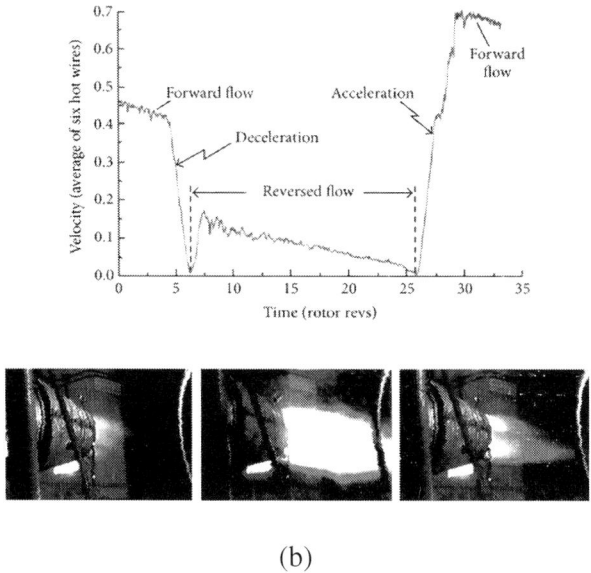

(b)

Figure 8: Showing (a) map of the compressor with information about dangerous threats [2, 3]; (b) deep surge cycle [3, 61]: the deceleration and acceleration of the flow in the compressor duct are a broadband impulse extortion for compressor blades and a bearing system (LCF concentrations) as well as with thermomechanical fatigue (TMF) for elements of a combustion chamber and the turbine.

Endurance problems of compressor blades can result also from new acoustic properties of a combustion chamber and the combustion process, incurred after the change of the fuel type, for example, using Jet-A1 or F-34 (NATO) in place of Jet-B or during high disturbance of temperature field in front of the turbine generated by carbon deposit of injectors [3].

Both old and new properties of disadvantageous extortions are unknown for the aero-engines user. Endurance problem of the compressor blades is noticed only after the first coincidence of cracking or breaking off the blade in service.

To sum up, fatigue problems of compressor blade are an effect rather than a cause. For the analysis of the problems, two approaches are used: classical and holistic. In the classic approach, the material fatigue results mainly from operating times (load cycles) and intensity of adverse phenomena (e.g., the human aspects and the specificity of aviation missions), Figure 9(a). In the holistic approach the material fatigue of compressor blade results from the level and the duration of disturbing the flow of the energy which aspects of the quality are affecting of the production, the repair and the use of the engine, Figure 9(b). The effective prevention requires applying the observer (the NDE and/or SHM method, methods of the signal analysis), which will be detecting not only the crack of blade, but also the cause of the hastened material fatigue (Table 1).

Table 1: Endurance threats of real blades permanence ("live") according to the theory of five elements [3]

Flow dynamics path	Kinematic loads path
Flow clocking	
Stall	Centrifugal force
Surge	Compressor nonaxial
Flutter	Rotor unbalance
Combustion instability	Compressor speed fluctuation
Foreign object in inlet	Structure resonances
Flight on big angles of the attack	Flight with big "g"

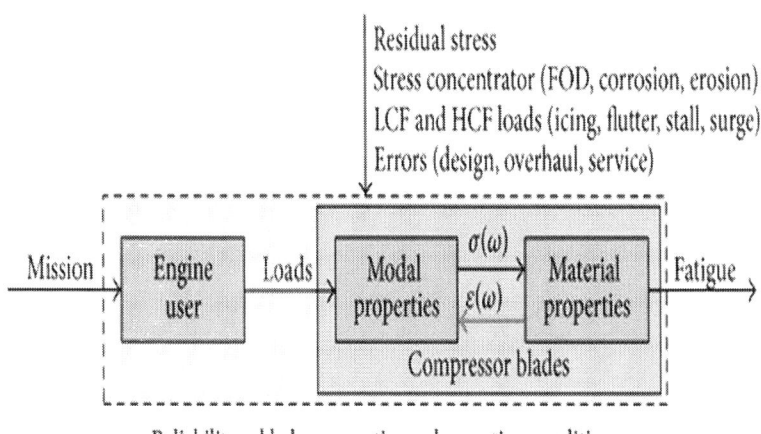

(a)

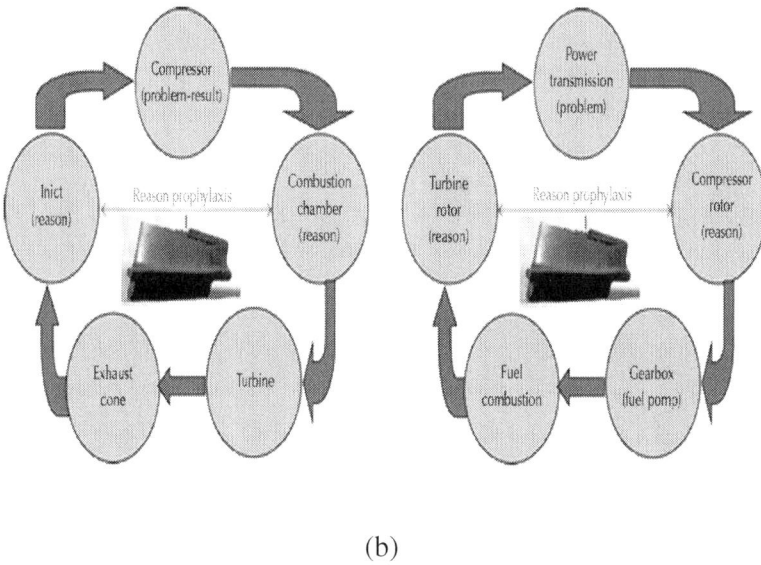

(b)

Figure 9: Analysis of cause and effect [3]: (a) classic approach with influence of human and operational factors on fatigue problems of the blades; (b) holistic approach—a compressor blade crack is a "result" of the wrong level of energy flow inside the engine. The "causes" must be sought at the inlet or the combustion chamber thermodynamic parameters (flow dynamics path) or mechanical power transmission quality and flight loads (kinematic loads path).

CONCEPTS OF EVALUATING BLADES

Effective and credible monitoring of the structure is possible, when technical problems are recognized, expected diagnostic symptoms are detected, and measuring and analytical methods (algorithms) could be applied.

Monitoring of the Vibration Magnitude

An intuitive diagnostic symptom of LCF problems is blades magnitude. Detecting the dangerous level of blade vibration during the work of an aeroengine and the change of flight conditions or the rotational speed of the engine are a base of the straightest preventive activity.

A report is describing the diagnostic rule

$$\text{if } A_{blades} < A_{max}, \text{ then OK else LCF threat.} \quad (1)$$

This approach has advantages and disadvantages.

Advantages

- Practicable methods in the service and flight (for a board monitoring system);
- Possible active control of blades LCF.

Disadvantages

- The level of the dangerous vibration amplitude is different for different mod (higher mod ⇒ smaller acceptance amplitude);
- The indicator of the dangerous vibration amplitude is not detecting the HCF and VHCF threat of blades;
- During the work of an engine, the vibration amplitude of the cracked blade little differs from healthy blades (influence of the centrifugal force).

Monitoring of the Mod Frequency

An intuitive diagnostic symptom of a blade crack is "a in change its modes frequency" (influence of active area change). The cracking propagation and blade break-off occur at limiting decrease in frequency [23]. The change of the blade frequency also results from erosion and the corrosion. Erosion is reducing mass of the blade and increasing the frequency. The corrosion is reducing the stiffness of the blade, and in the end the frequency of the blade is decreasing. Single compact FOD practically is not changing modal parameters of the blade, but it is a mechanical notch with concentrator of stresses.

A report is describing the diagnostic rule

if $f_{blade} \in \langle f_{min}, f_{max} \rangle$, then OK

else if $f_{blade} > f_{max}$ then

    Work hardening or Erosion

else if $f_{blade} \geq (f_{min} - \Delta f_{max})$ then

    Softening else cracking.

(2)

This approach has advantages and disadvantages.

Advantages

- Practicable methods in the overhaul and service of the aero-engines;
- Possible active control of blades fatigue;
- Possible comprehensive diagnostics of the engine. The blades of the compressor palisade are mechanical filter about known average parameters.

Disadvantages

- The frequency value of cracking blades depends on the crack position, parameters at the top of the crack edge (hardening or softening), and the loading. The modal frequency of the blade can suit different sizes of cracks (Figure 10).
- The blade frequency and mass distribution are tuned during assembly. Diagnostic symptom can be distorted during the manufacture and repair of compressor.
- Blades' frequency check offers too short prognosis horizon (Figure 5). It is sufficient in the blade health monitoring only; for example, in the tip-timing method which is used to detect dangerous blades vibration and open cracks during engine operation.

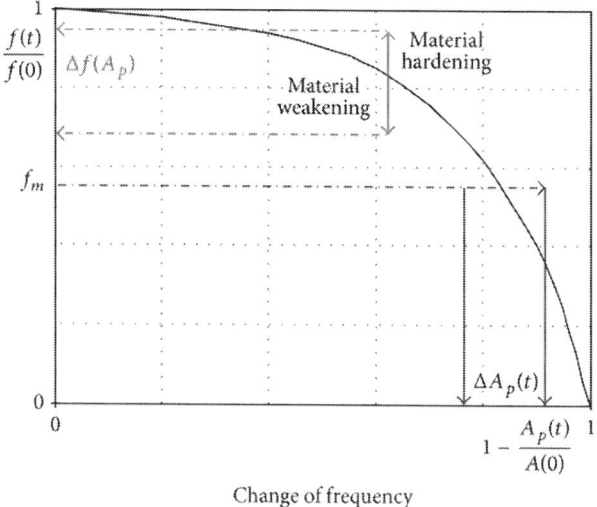

Figure 10: Lack of explicit copying of the size of the crack in the 1st modal frequency of blades [3].

Damage Monitoring of the 2nd Stage

More sensitive detection of fatigue is based on the relationship between changes in the material structure during 1st and 2nd stage of damage (before open crack), Figure 5, and subtle measurable changes of parameters (diagnostic symptoms).

Generally speaking, crystal lattice imperfections have a mechanical strengthening effect, since the lattice defects act as obstacles to the movement of dislocations when a mechanical stress is applied. Different strengthening mechanisms can be distinguished depending on the type of lattice defect contributing to the obstruction of moving dislocations [56, 64, 65]:

- solid-solution strengthening (interstitial/substitu-tional impurity atoms);
- strengthening from point defects (due to vacancies);
- work-hardening or strain-hardening (due to other dislocations);
- grain boundary strengthening;
- martensite strengthening (phase transformation);

- strengthening from fine particles (due to precipi-tates/inclusions).

Compressor blades work in temperature which is below 30% of melting temperature, that is, in cold working regime. Overload, stress concentration near nodal line, erosion and corrosion pitting, and fatigue, Table 2, mainly influence on the dislocation density p_d and the grain diameter d, defined as the average grain diameter. Cold working and fatigue change mechanical parameters of material (yield point σ_y, tensile strength σ_{utl}, hardness and microhardness, Young modulus, damping, and nonlinearity) [66–68]. In the end modal properties of the blade are changing. The fatigue damage progression can be divided into different (partially overlapping) stages, based on studies of the basic structural changes [56, 64–71].

Table 2: The fatigue of metal compressor blades at numbers of load cycles in the range 10^0–10^{12} [69]

Number of cycles	Description
10^0–10^3	Low cycle quasi-static or fatigue fracture (LCF problem) at availability of large "microplastic" deformation in some zones of failure. The crack nucleation below the surface.
10^3–10^5	Low cycle fatigue fracture (LCF problem) at availability rather small "macroplastic" deformation in a zone of failure (when $\sigma_e \leq \sigma_a \leq \sigma_y$, σ_e —a limit of elasticity). The crack nucleation from the surface.
10^5–10^8	Many classical cycles fatigue fracture (HCF problem) at availability of "microplastic" deformations in micro and macrovolumes near a zone of fatigue (when n $\sigma_a \leq \sigma_e$). The crack nucleation from the surface.
10^9–10^{12}	Fatigue fracture on super high bases (VHCF problem) at availability of "microplastic" deformation in microvolumes near a zone of failure (when n $\sigma_a \ll \sigma_e$). The crack nucleation on the soft inclusion below the surface.

The fatigue damage progression can be divided into different (partially overlapping) stages, based on studies of the basic structural changes [56, 64–71].

Fatigue Damage Initiation. Even at cyclic stress amplitudes below the macroscopic yield stress, the cyclic mechanical loading can plastically deform the material locally on a microscopically small scale: such microplastic flow first occurs in the grains that are stressed with the highest shear stress amplitude and near inherent material imperfections (inclusions, scratches, and voids). Indeed, tensile residual stress concentrations, associated with such imperfections, lower the actual applied stress at which the material starts to plastically deform locally in some individual grains. The initiation of damage due to cyclic deformation therefore consists of microstructural changes associated with localized micro-plastic deformation in some individual grains, that is, the development of slip bands, the generation of dislocations (increase of dislocation density), and the rearrangement of dislocations into dislocation tangles, dislocation walls, and persistent slip bands. These persistent slip bands can be envisaged as embryonic fatigue cracks.

Result. Nucleation of microsized cracks along the developed slip bands in a number of grains.

Slip Band (Stage-I) Crack Growth. Consider a microcrack that is initiated inside an individual grain of a polycrystalline material. Such microcrack can grow further under sufficiently applied cyclic stress, along slip planes of high shear stress. Then, to develop further, the crack must propagate into the neighbouring grains, which have different lattice orientations and therefore different slip systems. For small microcracks to propagate, the crack needs to reorientate at the grain boundary towards a particular slip direction of the surrounding grain. Typically, the majority of lifetime corresponds with microcrack (nucleation and growth), which is moreover a regime of stable damage progression.

Result. Formation of dominant crack(s) (with dimensions of typically a few to ten grain diameters wide).

Transgranular (Stage-II) Crack Growth. As the microcrack propagates, the plastic zone around the crack tip increases and the resistance to crack growth diminishes. The crack becomes insensitive to grain boundary obstacles and to the particular slip systems of individual grains: the crack now develops in the plane normal to the tensile stress direction, and at much faster rates per loading cycle compared to stage-I crack growth.

Result. Growth of a well-defined crack, along the plane normal to the applied stress direction, and coalescence of microcracks towards a macro-crack with such critical macroscopic dimensions, that the remaining cross-sectional area of the material can no longer support the maximum applied load, and the material fails by ultimate fracture during the last stress cycle.

The material's performance concerning the fatigue damage process is typically characterized by a S-N curve, also known as a Wöhler curve, which gives the cyclic stress amplitude σ_a as a function of the number of cycles to failure N_f, the latter in a logarithmic scale. Classically the S-N curve is not crossing 2.10^8 cycles what is insufficient for the compressor blades which TBO is copied through 10^{10} to 10^{14} cycles. Taking the VHCF risk into account requires using the widened S-N curve, appointed on the base of the bimodal theory of the metal fatigue (containing other bifurcations). About the real permanence of the blade (for the given level of exploitation stresses) a state of its surface (level of erosion and/or corrosion, FOD) is deciding (Figure 11).

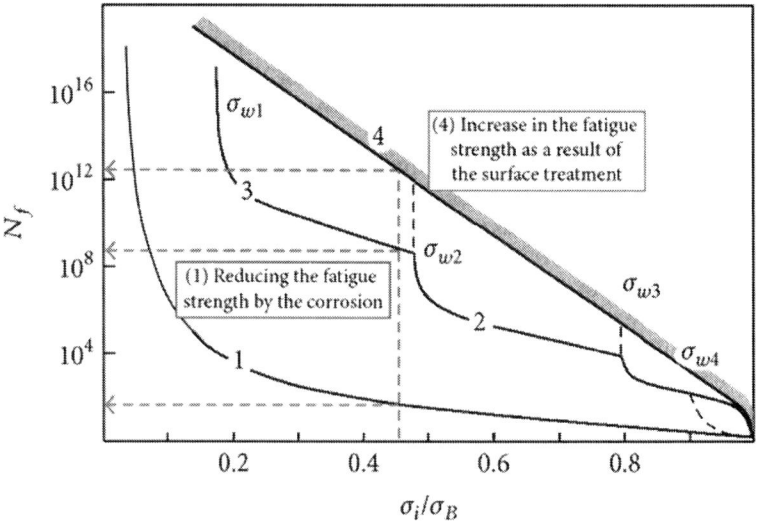

Figure 11: Influence of surface state on the number of cycles to blade failure N_f [3, 56].

Strengthening of a metal, which represents the increase in the resistance to yielding or plastic deformation, can be obtained by changes in microstructure that impede the motion of dislocations [67, 68, 71]. Based on the type of obstacles that hinder the motion of dislocation and hence increase the strength, the yield strength σ_y of steels is usually expressed in the form of generalised equation where the contribution of all the strengthening mechanisms is added as follows [72]:

$$\sigma_y = \sigma_0 + \sigma_{SS} + \sigma_P + \sigma_{GB} + \sigma_D + \sigma_T, \qquad (3)$$

where σ_0 is the lattice friction, σ_{SS} is the solid solution strengthening, σ_P is the precipitation strengthening, σ_{GB} is the grain boundary strengthening, σ_D is the dislocation strengthening, and σ_T is the texture strengthening.

In theoretical considerations, the Hall-Petch and the Bailey-Hirsch relations (4) between microstructure and mechanical parameters are given as follows:

$$\sigma_y \cong \sigma_f(T) + \frac{k_{HP}}{\sqrt{d}},$$

$$\sigma_y\left(\text{after } \varepsilon_p\right) = \sigma_0 + \tau_i \cong \sigma_0 + \alpha G b \sqrt{\rho_d} \qquad (4)$$

with σ_f being the friction Peierls-Nabarro stress required to move a dislocation in a single crystal, T being material temperature, k_{HP} being material-dependent Hall-Petch constant which represents the difficulty required to unlock or generate dislocations in neighbouring grains, d being the grain diameter, σ_0 being the lattice friction, τ_i being the shear internal stress, α being a constant, G being the shear modulus, b being the crystal lattice parameter (base length of cubic unit cell), and ρ_d being the dislocation density.

Typical values for carbon steels are

$\alpha \cong 0,4, b = 0.286 \text{nm}, d = 10^{-4} - 10^{-6} \text{m}, G = 80 \text{GPa}, k_{HP} = 0.74 \text{MPa}\sqrt{m}, \sigma_f = 70 \text{MPa}, \text{and } \sigma_0 = 100 \text{MPa}.$

The residual "life" of material ζ (relative residual time to blade break-off) is given by local level of dislocation density and the following relation:

$$\zeta = 1 - \frac{\sqrt{\rho_d} - \sqrt{\rho_{d0}}}{\sqrt{\rho_{d\,max}} - \sqrt{\rho_{d0}}} \quad (5)$$

with ρ_{d0} being the dislocation density for well-annealed materials, and $\rho_{d\,max}$ being the dislocation density for ductile strength. Typical values for carbon steels are $\rho_{d0} = 10^{-10}$ m^{-2}, $\rho_{d\,max} \cong 10^{-15}$ m^{-2}.

The second stage of blade damage can be observed, for example, in "a resonance curve," (RC) measured by a laser point head during modal frequency testing [3, 23]. This approach is sufficient to solve overhaul problem (level II) with experimental modal analysis method. A report is describing the diagnostic rule (6), which will be developed at the description of method:

if RC is for a linear object, then

OK or Work hardening

else Softening or Cracking. (6)

This approach has advantages and disadvantages.

Advantages

- Simple algorithms of the data analysis and diagnostic rules.
- High performance of examinations average time of objective testing (of parameters chosen mod and of the structural health condition) does not exceed 3 minutes per the blade.
- Possible reliable prognosis and reduce of the risk of wrong diagnosis during engine overhaul.
- The method is made available with base knowledge about relations between the material fatigue and modal properties which is being used by the tip timing method.

Disadvantages

- Required is direct access to the blades.
- The approach requires the selection of a new sensitive observers (measurement technics and analysis methods) as well as recognition and verification of new diagnostic symptoms.

Damage Monitoring of the 1st Stage

If the blades are made of ferromagnetic material, the biggest extension of the prognosis horizon is possible. The dislocation density, the diameter of the grain, the history of mechanical load (residual stress) and applied loads change not only mechanical parameters (3) but also the state of magnetizing blades, see Figure 12 and relation (7). Changes of material magnetizing, resulting from lattice-spin coupling (L-S) in the atomic scale, distribution of residual stress in the atomic micro- and macroscale, and magnetomechanical effects (reversible and irreversible) in the macroscopic scale, enable the detection of 1st phase of the damage [73–75]:

$$\mathbf{M} = \mathbf{M}_i + \mathbf{M}_r = (1 + k_H)(1 + k_\sigma)(1 + k_T)\mathbf{M}_0$$

$$H_c \propto \{\sqrt{\rho_d}, d^{-1}\} \longrightarrow \{L_i^{-1}, \sigma_r\}$$

$$B_r, \mu_{r\,\text{max}} \propto \{\rho_d^{-1}, d\}$$

$$\mu_i \propto \{L_i^2, \sigma_r^{-1}\} \qquad (7)$$

with M being magnetization; M_i being the induction magnetisation; M_r being the residual magnetisation; M_0 being initial state of magnetizing the blade; k_H, k_σ, k_T being appropriately influence of external field, stress, and temperature; H_c being the coercive force; B_r being the remanence; $\mu_{r\,\text{max}}$ being the maximum magnetic permeability; μ_i being the initial magnetic permeability; σ_r being the magnitude of unidirectional internal stress which represents the irregularly fluctuating magnetoelastic energy distribution; L_i being the periodic distance between internal stress centre [76]

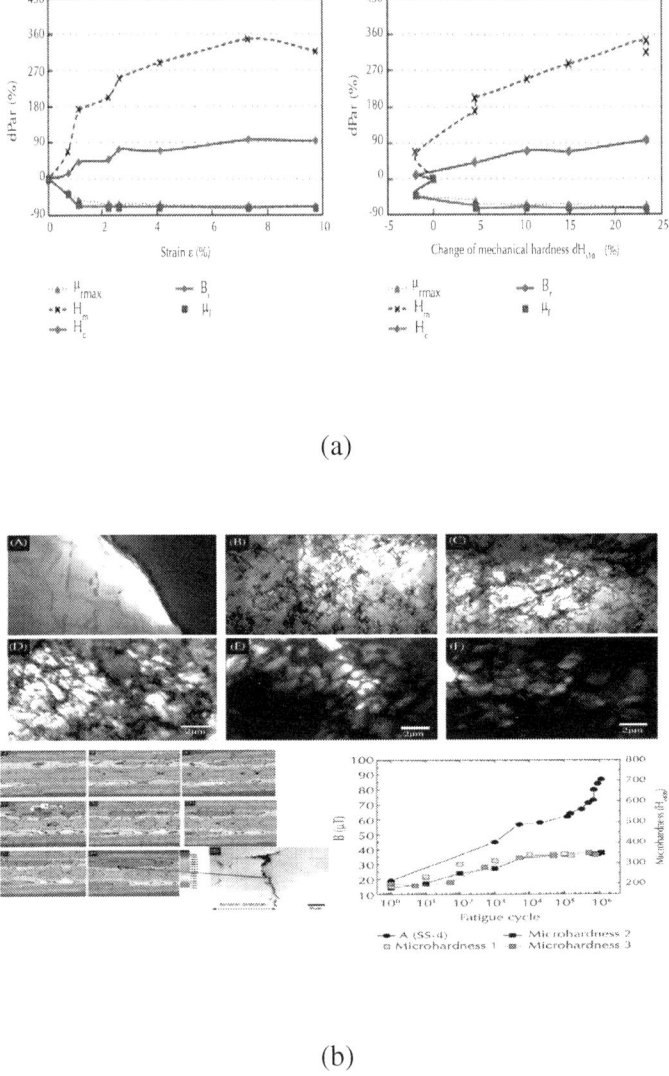

(a)

(b)

Figure 12: Showing (a) influence of static strain hardening on magnetic parameters low carbon steel with 0,17% C ($H_m = H(\mu_{r\,max})$), $d\mathrm{Par} = (\mathrm{Par}(\varepsilon) - \mathrm{Par}(\varepsilon = 0)/\mathrm{Par}(\varepsilon = 0))$ [73]; (b) influence of HCF on TEM microstructure of fatigue 430 stainless steel (after 0 cycles, 1×10^3 cycles, 1×10^4 cycles, 2×10^4 cycles, 4×10^5 cycles, 5×10^5 cycles) and remanence magnetization map of the steel (after: (A) 0 cycles, (B) 1×10^3 cycles; (C) 5×10^3 cycles; (D) 2×10^4 cycles; (E) $1,2 \times 10^5$ cycles;

(F) $1{,}45\times 10^5$ cycles; (G) $2{,}8\times 10^5$ cycles; (H) $9{,}2\times 10^4$ cycles. SQUID scanning area 10 × 20 mm) [74].

Magnetic properties of the blades are dependent on the microstructural type, of additions alloy and residual stresses, as well as level of material damage. A report is describing the diagnostic rule (8), which will be developed at the description of method:

$$\text{if } \mathbf{B} \in \langle \mathbf{B}_{\min}, \mathbf{B}_{\max} \rangle, \quad \left|\frac{dB}{dx}\right| < \left|\frac{dB}{dx}\right|_{\max}$$

then OK

else Hardening, Erosion, Softening,

Cracking, Stress Concentration Zone. (8)

This approach has advantages and disadvantages.

Advantages

- Detection of reversible changes of the material fatigue (1st phase of damage).
- Possible of remote observation of magnetizing blades through paramagnetic casing of compressor. It is sufficient to solve service problem—detect fatigue risk before the first blade crack (level I).
- Observation of magnetizing blades during the engine stop (using irreversible magnetomechanical effects), the small rotation speed (from zero rpm), and the engine work on the operation range.
- Possible reliable prognosis and reduce of the risk of wrong diagnosis during engine overhaul and service.
- Possible analysis postfactum of damage elements and identification of load condition prevailing during the initiation and the propagation of the crack.(vi)The solution can be used for diagnosing other ferromagnetic elements of the plane, for example, of bearing, gears, shafts, pressure vessels, and landing gear.(vii)The method is made available with base knowledge about relations between the material fatigue and stress-strain inducted magnetization and symptoms using by the experimental modal analysis and the tip timing methods.

Disadvantages

- Strong nonlinear rules describing the magnetization of the ferromagnetic parts, particularly in the weak magnetic field.
- The approach requires the selection of a new sensitive observers (measurement technics and analysis methods) as well as recongnition and verification of new diagnostic symptoms.
- Only for ferromagnetic materials and some paramagnetic steel.

SHM AND NDE METHODS

Magnetic and magnetomechanical properties of ferromagnetic blades are changing their mechanical and modal properties. The state of magnetizing blades also affects the tip timing signal (when there is an induction, vary reluctance or eddy current sensor is used to detect moving blades). The theory and experience of experimental modal analysis are an input to the tip timing method. For paramagnetic blades which material is not showing of phase transformation under the influence of stresses, the first method (Metal Magnetic Memory method) is not applicable.

The Metal Magnetic Memory Method

The MMM method (only NDE method according to ISO 24497:1-3 (2007)) is based on three pillars:

- magnetomechanical effects existing in ferro-magnetic material being located in a weak magnetic field of the Earth which was described by theory of the micromagnetism [76–88];
- magnetovision: the remote passive observation magnetic field near the testing element;
- magnetostatic: solving the opposite issue of magnetostatic in the destination of the magnetizing trend from magnetic anomaly (symptom of defects, stress concentration zone, and change of element shape).

The MMM method is a typical passive observer of the signal analysis which is possible to apply both to NDE and SHM applications. Signal analysis is the process of determining the response of a system, due to some generally unknown excitation, and of presenting it in a manner which is easy to interpret (Figure 13).

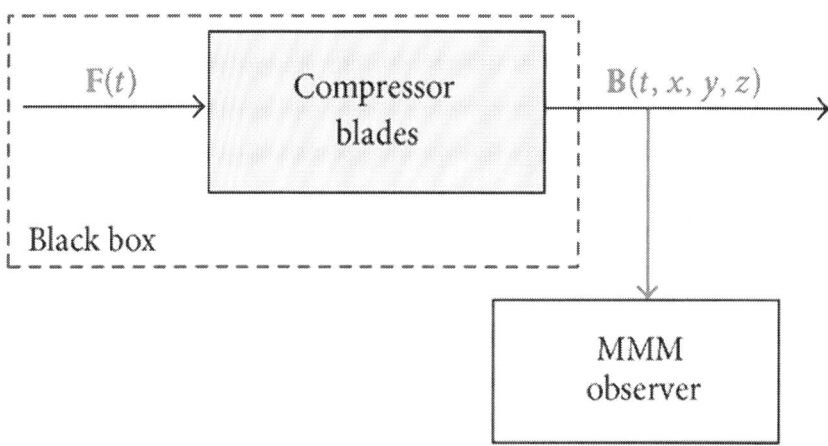

Figure 13: Signal analysis of the blade magnetizing (magnetic induction B in the close of the its surface) which the health monitoring and load history of the blade described.

In the macroscopic scale a constitutive law (9) is copying magnetic properties of the blade:

$$\mathbf{B} = \mu \mathbf{H} = \mu_0 (\mathbf{H} + \mathbf{M}) \tag{9}$$

with μ_0 being magnetic permeability of vacuum (in SI units system $\mu_0 = 4\pi \cdot 10^{-7}$ H/m); M being material magnetization [A/m], H being external magnetic field [A/m], and B being magnetic induction (magnetic flux density) [T].

Reversible Magnetomechanical Effects

Apart from the constitutive law B(H) or M(H), there is a second class of macroscopic observations that needs to be introduced. It is observed that when a ferromagnetic specimen is subjected to a magnetic field, its magnetization as well as its length change—Joule effect, which is described by tensor rule (10). When a ferromagnetic specimen is subjected to a mechanical stress, both it is length as well as it is magnetization change—Villari effect, which is described by tensor rule (11) and depicted in Figure 14. The actual distribution of material magnetization (molecular currents in the material) can be observed indirectly by measuring the magnetic field distribution in the nearby the object.

$$\varepsilon_{ij} = s_{ijkl}^{HT}\sigma_{kl} + d_{ijn}H_n, \tag{10}$$

$$B_m = d^*_{mij}\sigma_{ij} + \mu_{mn}^{T\sigma}H_n, \tag{11}$$

where $d = \partial\varepsilon/\partial H|_\sigma$ and $d^* = \partial B/\partial\sigma|_H$ are magnetomechanical coefficients which are appointed experimentally for given material at the constant tensile (stresses) σ or constant magnetic field H [75].

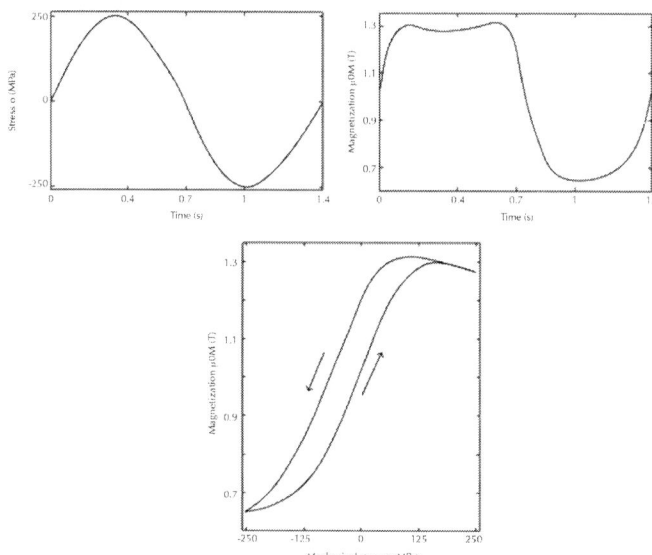

Figure 14: Figure 14: Stress inducted magnetization in low carbon steel (R_e = 390 MPa, stress level below yield strength, static magnetic field H = 800A/m) [76].

In (10) and (11) an influence of the change of material temperature and losses of the internal energy were omitted. Equations are describing only "reversible magnetomechanical effects."

Zones RSC of local residual stress concentration, plastic, material anisotropy (mechanical and magnetic) and dislocation concentration are potential place of cracking nucleation and local magnetic anomaly [4, 72, 76–79]. Influence of the local plastic strain of material (LCF, HCF and VHCF problems—Table 2) the best is visible in the weak magnetic field [80]. "The passive magnetic observer of the blade health monitoring (e.g., the metal magnetic memory method) is favoured."

Magnetomechanical Damping

Applying a stress to a ferromagnetic blade causes a variation of magnetization due to the magnetoelastic coupling, which results in the so-called "ΔE effect" and also in a related dissipation of mechanical energy during loading/unloading or in case of vibration. The latter effect can give rise to a strong magnetomechanical damping with stress-dependent and stress-independent components [81].

Experiments show that ferromagnetic materials have a higher internal friction than the paramagnetic and diamagnetic because of phenomena of an electromagnetic nature resulting from the application of elastic fields. Considering five main contributions to the total energy of a ferromagnetic material without an external field (exchange energy W_{ex}, magnetocrystalline anisotropy energy W_k, magnetoelastic (or magnetostrictive) energy W_λ, magnetostatic energy W_m, and energy of magnetic domain walls W_w), four main mechanisms of magnetomechanical damping may be defined:

- magnetoelastic hysteresis damping Q_h^{-1}
- macroeddy-current damping Q_a^{-1}
- microeddy-current damping Q_u^{-1}
- damping at magnetic transformation Q_{PhT}^{-1}. Therefore, the total magnetomechanical damping Q_m^{-1} in ferromagnetic blade can be considered as sum of these components:

$$Q_m^{-1}(\varepsilon, \omega, T) = Q_h^{-1}(\varepsilon, \omega, T) + Q_a^{-1}(\omega, T)$$
$$+ Q_\mu^{-1}(\omega, T) + Q_{PhT}^{-1} \qquad (12)$$

contrary to Q_a^{-1} and Q_u^{-1}, the hysteretic contribution Q_h^{-1} depends on the strain amplitude. The damping Q_m^{-1} is also dependent on the load frequency ω, material temperature, and initial conditions (micro- and macrostructure, magnetization, and residual stress). "The Q_m^{-1} is nonlinear."

Irreversible Magnetomechanical Effects

Losses of the internal energy are being observed in the weak DC magnetic field in the form of

- the magnetization hysteresis loop (Figure 15);
- growing magnetizing material under the influence of the cyclical load (LCF an HCF fatigue) [82]. Observed change of magnetizing material depends on the level of stresses and the number of cycles;
- change of magnetizing material after unloaded ("the metal magnetic memory" or "the first loading/ unloading effect") [77, 83]. The ferromagnetic blade has feature of the strain gauge with the memory of the maximum load.

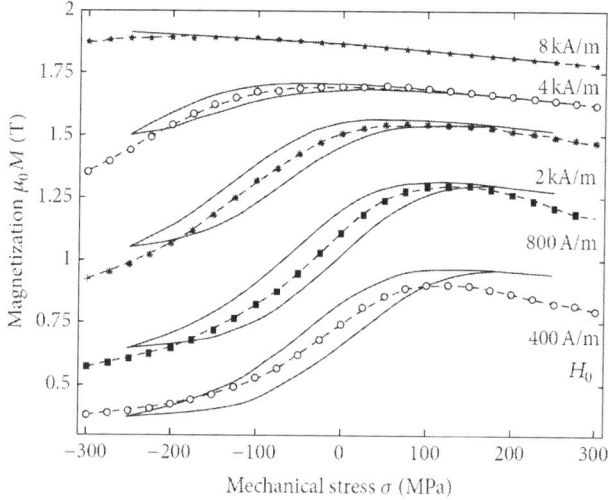

Figure 15: Experimentally obtained magnetomechanical hysteresis loop M(σ(t)) for different setting of external DC magnetic field H_0 (full lines) and compared to the reversible anhysteretic magnetomechanical behaviour at corresponding setting of H0 (dashed line with symbols). Material: low carbon steel with 0,12% C [76].

Measuring Equipment

Potential possibilities of the MMM method were tested in active and passive experiments, which used a compass (simple magnetometer), GM-04 Magnaflux magnetometer with Hall sensor [89], 3D MEMS anisotropic magnetoresistive sensor (HMC 5843 Honeywell demo board) [90], and Energodiagnostika TSC-1 M-4 recorder with multichannel transduction sensors (scanning devices) [91]. The Earth's magnetic field (Bf ≅ 50 μT) and electromagnetic noise are natural source of the external magnetic field.

The Experimental Modal Analysis Method

Experimental modal analysis (a tool of structural analysis) is an effective aid in solving blades' fatigue problems. It allows finding an answer to the question: "Why does a blade crack?", not only: "Is it cracked?".

The modal parameters of all the analysis modes (within the frequency range of interest) constitute a complete dynamic description of the blade structure [3, 17–21]:

- material,
- geometry,
- the influence of surface treatment and adding protection coating,
- technical health (structural heterogeneity, crack, and fatigue).

The characteristic feature of blade vibration measurement on a modal excitation system is knowledge of both a force level and a blade response on it (Figure 16(a)). This can be done by stimulating the system withmeasurable force and studying the response/force ratio. For linear system this ratio is an independent, inherent property which remains the same whether the system is excited or at rest. That is why it is possible to identify blade modal properties for following modes.

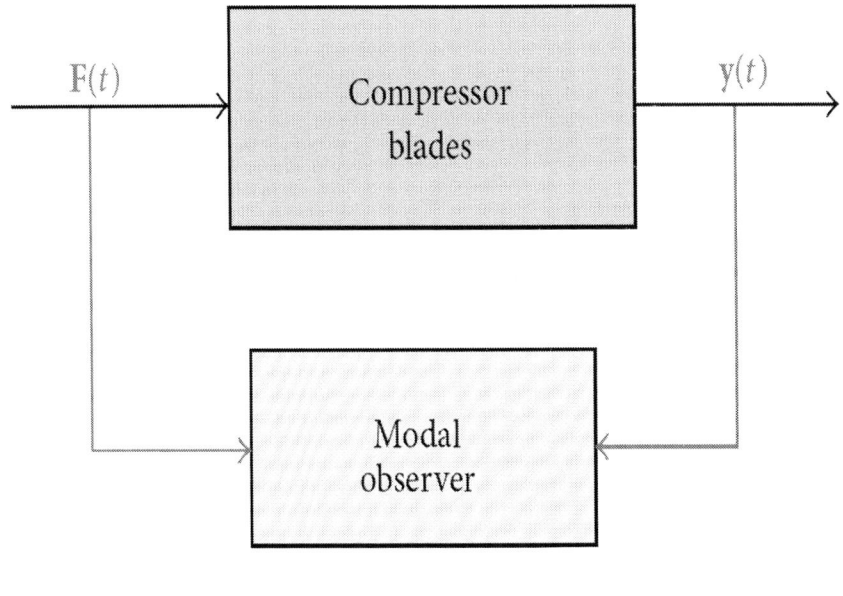

(a)

High Sensitive Methods for Health Monitoring of Compressor Blades...

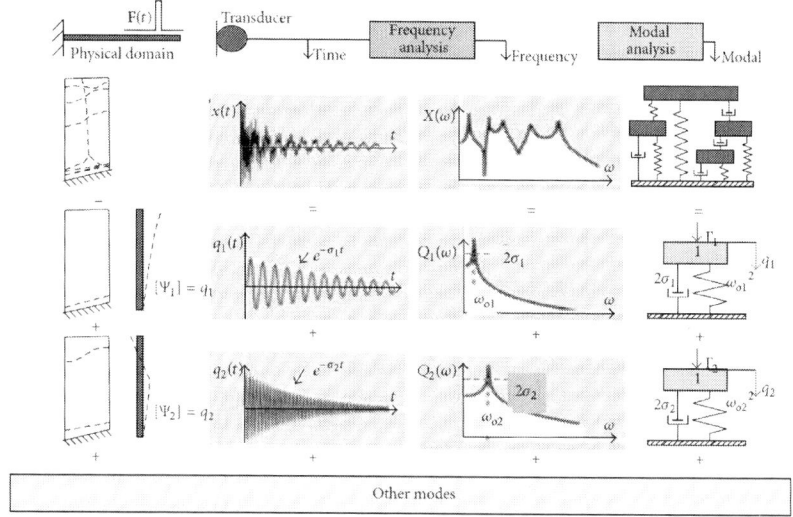

(b)

Figure 16: Blade health and mechanical properties analysis using the experimental modal analysis.

Structural response of the compressor blade (a lightly damped structure) can be represented in different domains. The modal description relates to descriptions in the spatial, time, and frequency domain (Figure 16 (b)).

In the illustration, each column shows the response of the blade after short struck represented in different domains.

Physical Domain: the complex geometrical deflection pattern of the blade can be represented by a set of simpler, independent deflection patterns, or mode shapes.

Time Domain: the vibration response of the blade is shown as a time history, which can be represented by a set of a decaying sinusoids.

Frequency Domain: analysis of the time signal gives us a spectrum containing a series of peaks, shown below as a set of SDOF (single-degree-of freedom) response spectra.

Modal Domain: we see the response of the blade as modal model constructed from a set of SDOF models. Since a mode shape is pattern of movement for all the points on the structure at a modal frequency, a single model coordinate q can be used to represent the entire movement contribution of each mode. The SDOF model is associated with a frequency, a clamping, and a mode shape. An important property of modes is that any forced or free dynamics response of structure can be reduced to a discrete set of modes. The modal parameters are as follows:

- modal frequency;
- modal damping;
- mode shape.

Measuring Equipment

The broadband identification (up to 20 kHz) of modal properties of a compressor blade, made of 18H2N4WA steel and Ti5.8Al-3.7Mo titanium alloy, has been made on the PSV-400 Polytec scanning vibrometer [92] and low power PZT exciter (Figure 17).

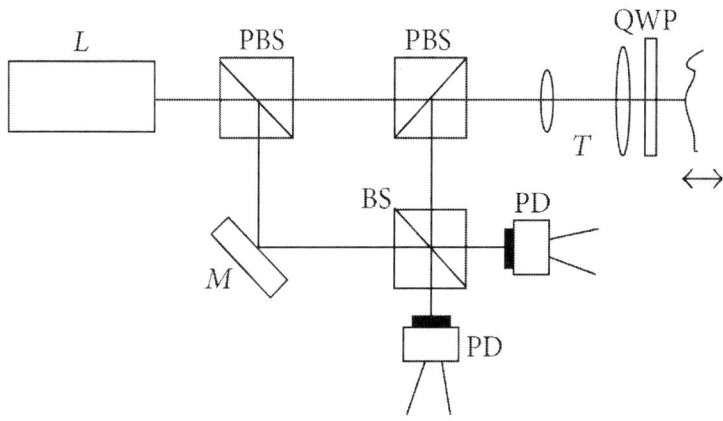

(a)

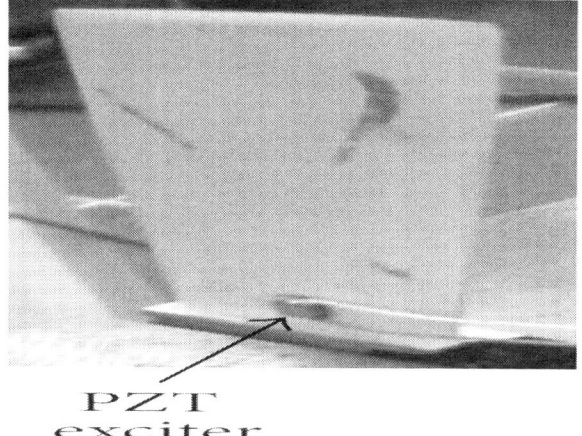

(b)

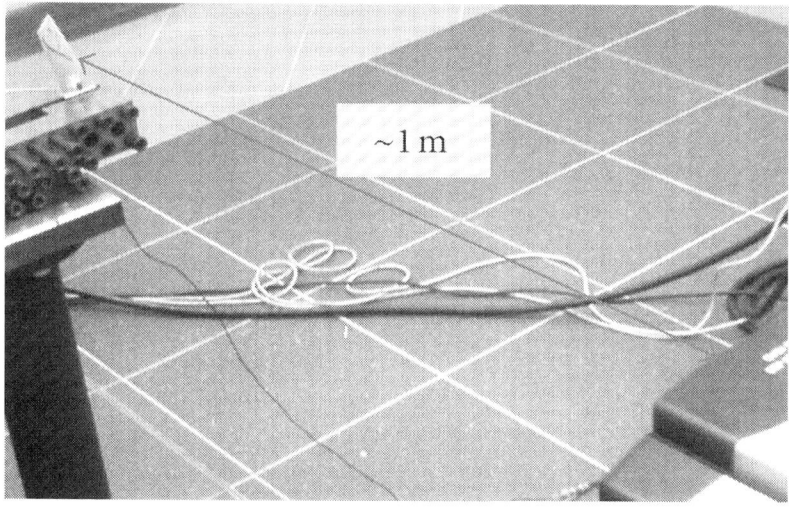

©

Figure 17: Experimental stand for the broadband (up to 20 kHz) modal identification of the compressor blade with the use of scanning laser vibrometer and PZT exciter.

This approach has advantages and disadvantages.

Advantages

- Automation of the measurement;
- Mode frequency and node lines are very precisely identified;
- Weak clamping of blade is possible;
- Input data to numerical modeling of modal and fatigue properties;
- Vibration frequency up to 20 MHz and vibration velocities up to 20 m/s;
- Scan area ($\pm 20°$ about X, Y) and grid definition;
- High speed (>50 points/s) and resolution (<nm, $0.002°$);
- High angular stability ($0.01°$/h).

Disadvantages

- High cost and weight measurement equipment (7,5 kg);
- The small vibration amplitude is not opening the crack.
- The identification of early fatigue and cracking symptoms of these blades has been made on the Brül & Kjær electro-dynamic exciter 4802T [93]. The experimental stand (Figure 18), used during the SO-3 and TW3-117 engine overhauls, included the following:
- the MTI Instrument laser measurement system MicroTrack II with CMOS measurement head LTC-120-40 [94];
- the Vibration Research Corporation VR-8500 controller that includes 24 bit A/D and D/A converters, and RISC processor [95];
- the Vibration Research Corporation Vibration View software to control the exciter, data acquisition, and analysis [96].

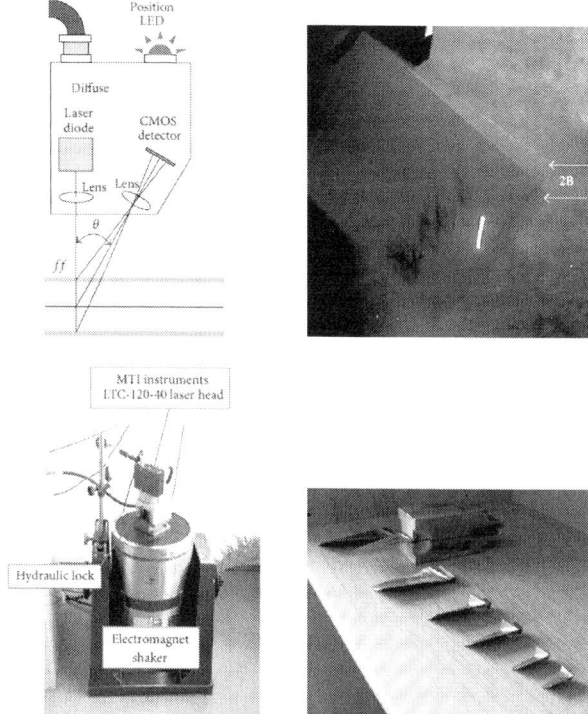

Figure 18: Workstation for the low-frequency (up to 4 kHz) modal identification of the compressor blades.

The sensitivity of measurement system is 100 mV/mm.

This approach has advantages and disadvantages.

Advantages

- Load up to 60 g ($a = 588$ m/s^2)—LCF & HCF test is possible;
- Low cost of laser head;
- Frequency response: 20 kHz max (laser) and 4 kHz max (exciter);
- Resolution at 20 kHz filter: ±5 μm; (v) Filter setting: 20 kHz–0.1 Hz;
- High temperature stability (0.005%/K) and linearity (0.05% FSR or better).

Disadvantages

- Single point measurement;
- Great demand of the electric power.

The Tip Timing Method

The tip timing idea consists in observing displacement of loaded component part, Table 3, with "irregular sampling" (one time to the turnover of the rotor). In our case, it will be a rotating and vibrating compressor blades. Blade vibration and deflection are a source of a time-interval change between flexible key phases. The tip timing observer (sensor) is built onto a fixed part of compressor. Its analog signal depends on the sensor type (Figure 19). Time period signal (time of blade arrival, TOA) would be measured with a frequency method or delay line (time-to digital converter) [3, 25, 96, 97].

Table 3: Potential field of tip timing method use

Object	Transmitter of putting the angle turnover
Wind turbine	Blade
Helicopter	Main rotor blade Tail rotor blade
Propeller aircraft	Propeller blade
Power turbine	Turbine blade
Jet engine	Compressor blade Turbine blade
Turboprop	Compressor blade Turbine blade
ABS systems, gearbox	Gear wheel
engine control unit	Gear wheel
bearings	Rolling element

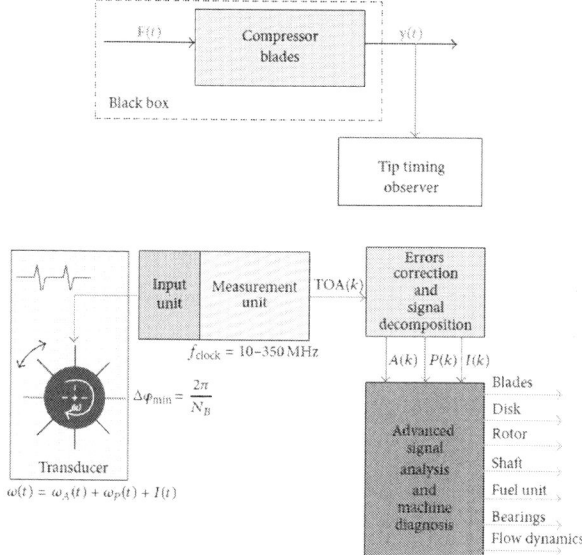

(a)

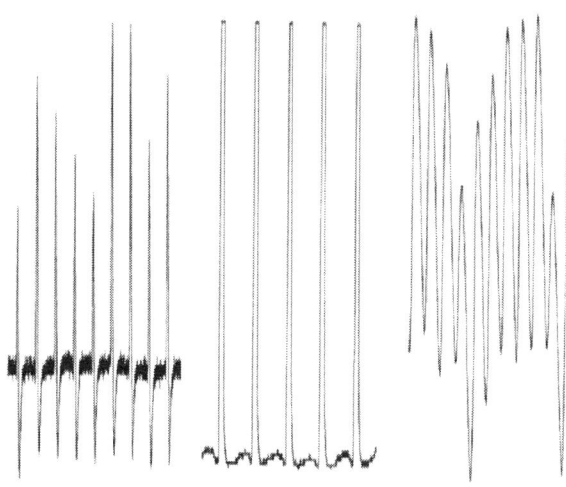

(b)

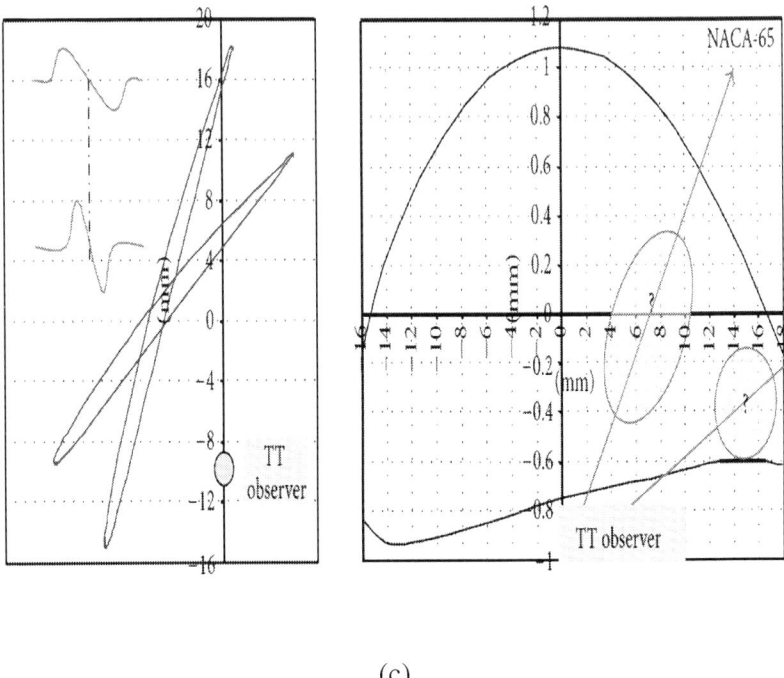

(c)

Figure 19: Showing (a) an idea and a block structure of the tip timing method; (b) a shape of analog signal for vary reluctance (VR), optical and eddy current (EC) sensor. Signal with VR and EC sensors also contains information about magnetizing blades; (c) main problem of VR and EC sensors—precise relation of the analog signal with putting the top of the blade with regard to the sensor (Where is the blade? Which point of the blade top is in the relation with the characteristic point of the analog signal?) [25].

Measured signal TOA(k) is discreet representatives of the continuous signal $S(t)$ which contains the following:

- aperiodic part $A(t)$—average instantaneous rotational speed of perfect stiff rotor;
- oscillating part $P(t)$—resultant from pitch errors, blade, rotor and disk vibration, and instantaneous rotational speed perturbations (from the engine control system, flow, g-force, clearance in a kinematic system, and torsional vibration);
- noise and weak oscillating components $I(t)$.

Signal S is described by the following relation:

$$S(t) = A(t) + P(t) + I(t) \tag{13}$$

so it is possible to design a general-purpose observer for real operating conditions of rotating parts and have a complex view on the following:
- disadvantageous dynamic phenomena (flutter, stall, surge, resonance, and load coupling);
- the influence of production, overhaul, and maintenance real conditions on the level of malfunctioning and fatigue prognosis.

Every component of S(t) is used to diagnose. An oscillating part P(t) is a main carrier of diagnostic information about blades damage and danger dynamics phenomena. Aperiodic part A(t) and part I(t) give the capability to compare new diagnostic symptoms to the health of machinery. Signal TOA(k)—a number of pulses $Code_i$ with clock frequency counting between key phases (blades)—includes "three groups of variables" (14) to be identified in effect of further numerical signal analysis

$$\begin{aligned} TOA(k) = Code_i &= K_{i,i+1} \text{Trunc} \left(\frac{t_{i,i+1}}{t_{clock}} \right) \\ &= \left(\frac{1+\zeta_B}{1+\zeta_\omega} TOA_{avg} \right)_{i,i+1}, \quad i \in \langle 1, 2, \ldots, N_B \rangle \\ TOA_{avg} &= \frac{2\pi/N_B}{\omega} \end{aligned} \tag{14}$$

with k being discrete time, $K_{i,i+1}$ the error and disturbance factor ($K_{i,i+1} = 1$ for data without error), N_B the number of blades, $t_{i,i+1}$ the time interval between two blade passes (with momentary pitch), t_{clock} the time period of generator pulses (with model of the time), TOA_{avg} the average momentary time of arrival of the perfect rotor and the blade palisade (without influence of rotor unbalance and vibration and blade pitch errors), ζB the jitter of blades group components, ζ_ω the jitter of rotor group components, and ω the angular velocity of ideal rotor

The jitter of blade group components includes

ζ_p: pitch errors (N_B of aperiodic variables);

ζ_{Bi}: vibration of i the blade (N_B of independent multimodal generators);

ζ_C: compressor case vibration;

ζ_{DP}: dynamic phenomena of TTM sensor

so it is described by the following relation:

$$\zeta_B(k) = \zeta_P(k) + \zeta_{Bi}(k) + \zeta_K(k) + \zeta_{DP}(k). \tag{15}$$

The jitter of rotor group components includes

ζ_F: influence of control unit and changes of the momentary rotation speed;

ζ_R: transverse and torsional vibration of rotor;

ζ_E: alignment error (eccentricity)

ζ_A: alignment error (misalignment), so it is described by the following relation:

$$\zeta_\omega(k) = \zeta_F(k) + \zeta_R(k) + \zeta_E(k) + \zeta_A(k). \tag{16}$$

The jitter ζ_ω is a source of FM modulation which is mainly problem of TOA(k) signal disintegration (to components A, P, and I after signal verification and numerical correction) and analysis of the blades health and vibration. Signal components of TOA(k) are obtained with the DETREND procedure (Figure 20).

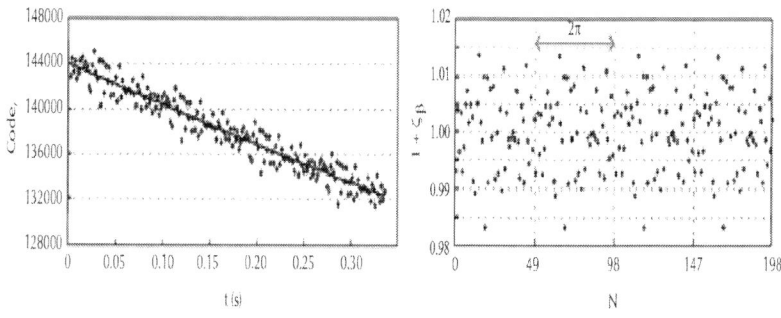

(a)

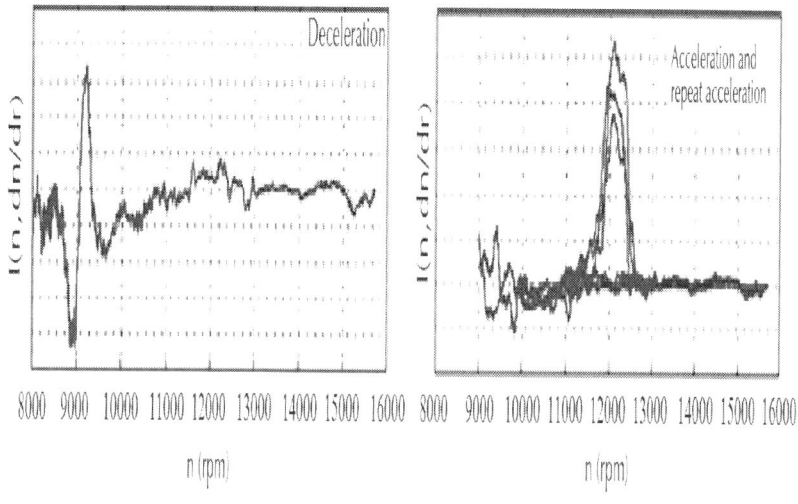

(b)

Figure 20: Result of the DETREND procedure for TOA(k) signal measured for (a) the last stage of SO-3 engine compressor (49 blades in the palisade)—the pitch errors are dominating in the jitter ζ_B; (b) the first stage of SO-3 engine compressor—the component $I(k) = I(n, dn/dt)$ is revealing the influence of some asynchronous parts of jitter ζ_ω on the component $A(k)$ (e.g., rotor vibration, fluctuation in the rotational speed on the compressor limits) [3].

The scope of interest of numerical data processing includes [3, 25–33]

- vibration level of all blades at the same time,
- disadvantageous dynamic phenomena,
- blade stress and health,
- disk health,
- engine health (the engine fuel system and the bearing system).

Signal subcomponents are obtained with the narrow-band filtering, AM/FM demodulation and spectrum analysis (e.g., CORDIC, DFT and DASP algorithms are used). Blade vibrations are shown in the form of phase distributions as points of phase trajectory crossing the phase plane [25, 55] (Figure 21).

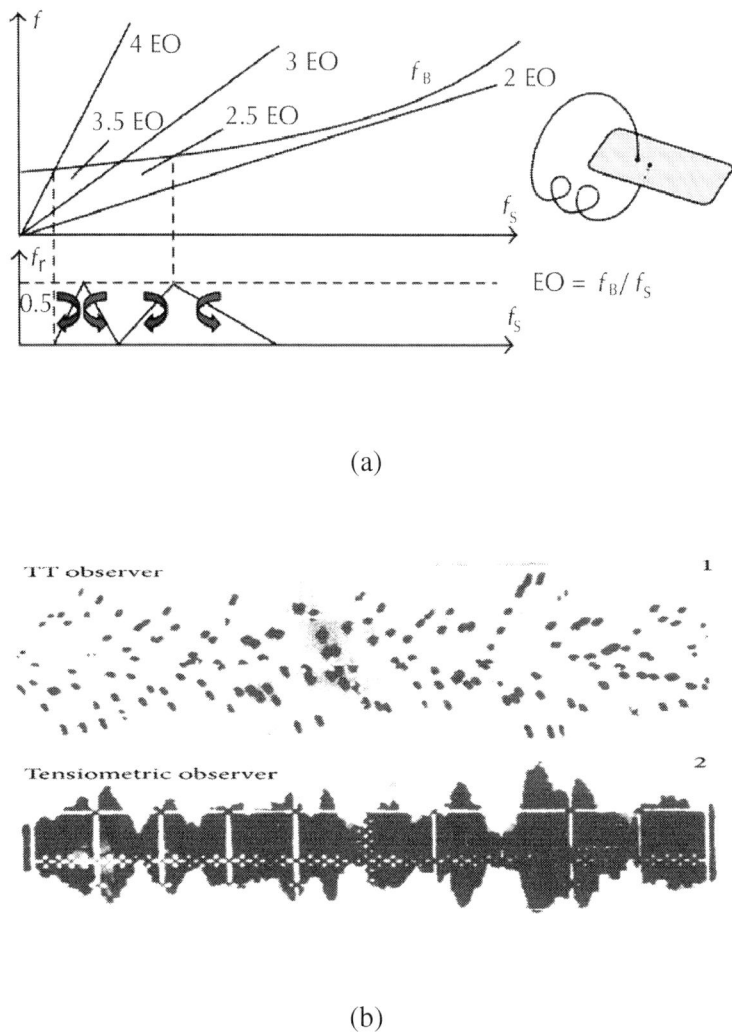

Figure 21: Relation between vibration of the blade top (the TT observer, $f$sampling $\cong f$rotor $\pm \Delta f$jitter) and stresses at the base of the one (the strain gauge observer, $f$sampling=4 kHz). Three first modal frequencies of the blade: 350 Hz, 1380 Hz, and 1890 Hz, max. rotational frequency of the rotor $f_{rotor,max}$ = 260 Hz [55].

A main characteristic feature of the tip timing method is information that lasts about a total number of modal frequency periods between two subsequent points of phase trajectory crossing the phase plane, with basic modal parameters of the blades preserved. This phenomenon enables detection of the LCF and HCF crack initiation and propagation in the blade during the engine operation. Other characteristic features of the TTM are the following.

- Irregular signal sampling rate: the Nyquist-Landau law describes discrete-time information.
- Periodic measurement data structure: data can be illustrated with matrix with N_B columns (number of blades) and rows that represent each full 360 degrees cycle of a rotor.
- The inherent in a signal oscillating parts that are not connected with blade vibration: there are two groups of oscillating parts of a signal: synchronized and nonsynchronized with rotor rotational frequency.

The SNDŁ-1b/SPŁ-2b System

In 1993, a diagnostic system was developed and introduced into the service on the TS-11 "Iskra" trainer. The system is based on results of the active and passive tests. It consists of [62] the following.

- The blade excessive vibration warning device SNDŁ-1b: a two-channel analogue phase detector that warns a pilot of conditions that can induce accelerated HCF of blades, for example, deposition of foreign matter such as ice, bird, or other resonance-based phenomena.
- The ground-based inspection instrument SPŁ-2b (digital phase detector, f_{clock} = 10 MHz) for
    - periodic recording of blade vibration,
    - inspections of the SNDŁ-1b health, with no need to have it disassembled,
    - detection of errors of the engine rotational speed indicators (in cabins I and II), with no need to disassemble them.
- The SPŁ-2b software: a set of programs that form the nucleus of the advisory/expert system used to diagnose the SO-3 engine. The software comprises
    - the database with text data: that is, a verified set of information on the object under examination and technical specifications of operating/monitoring it;

- the database containing measuring data: more than 7000 records on the Polish population of the SO-3 engines, collected during more than 20 years, taken at OAT = 248 to 308 K (−25 to +35º C), p_A = 959,9 to 1026,6 hPa (720 to 770 mm Hg), and humidity of 20 to 100%; also, the SO-3 overhaul-delivered data (collected during 15 years), including, among other things, information on frequency spacing of blades in the blade ring. The base also includes hidden software usage files (logs).They are made automatically without user knowledge. They are the base to valuate diagnostics system usage correctness;
- the database containing numerical models: compressor blades, the fuel system, and the rotor bearing system. The models are also used to identify how errors made during manufacturing, operation, and repairs/overhauls can affect the engine operational safety. Another application can be the postfactum analyses of air accidents;
- the database with diagnostic rules: contains the algorithms used to interpret measuring results and bringing diagnosis of the first-stage compressor blades and the transmission, as well as the expert diagnostic of a fuel system. The diagnostic rules also facilitate automatic identification and verification of the source of measurement data (the engine type and the serial number thereof) and software-based synchronisation of taking subsequent records. The database comprises also procedures of identification and correction of measurement errors and procedures of identification of errors in the operational use of the diagnostic system.

RESEARCH RESULTS

Chosen findings were obtained during active and passive experiments.

The Metal Magnetic Memory Method

Very good relation has been observed between the MMM results and blade node lines after LCF tests [3]. Local magnetic anomaly has been also observed near the close crack gap after HCF tests (Figure 22). The MMM method is widening the possibility of the blade verification by the RSC detection before the appearance of the endurance fracture (Figure 23). Nevertheless, the most interesting phenomenon is nondestructive detect of blade erosion and stress prehistory (a change of residual magnetization) after the engine stopped (Figure 24). Based on previous research and theoretical evidence does not rule out the possibility of diagnosing VHCF problems by MMM method. During measurements of the state of blades magnetizing through the compressor casing new symptoms was demonstrated for the tip timing method (Figure 25).

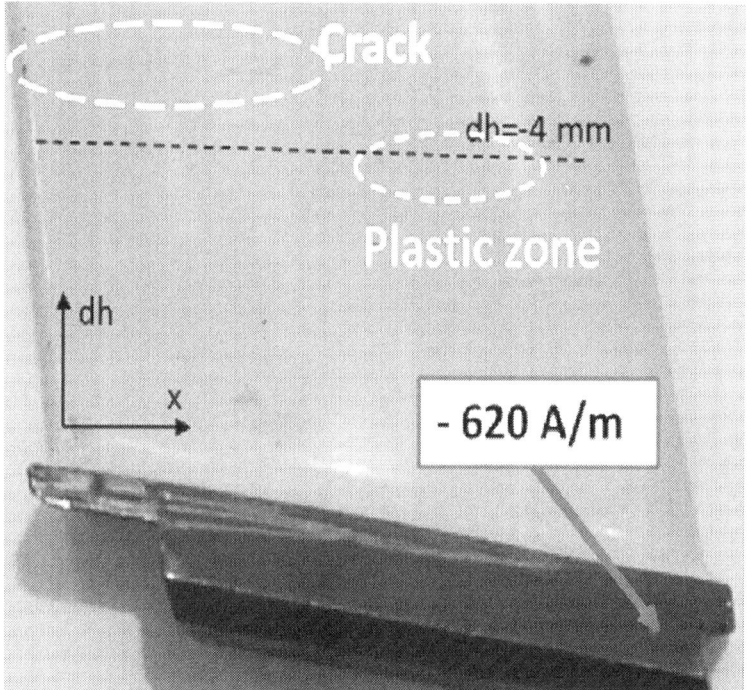

(a)

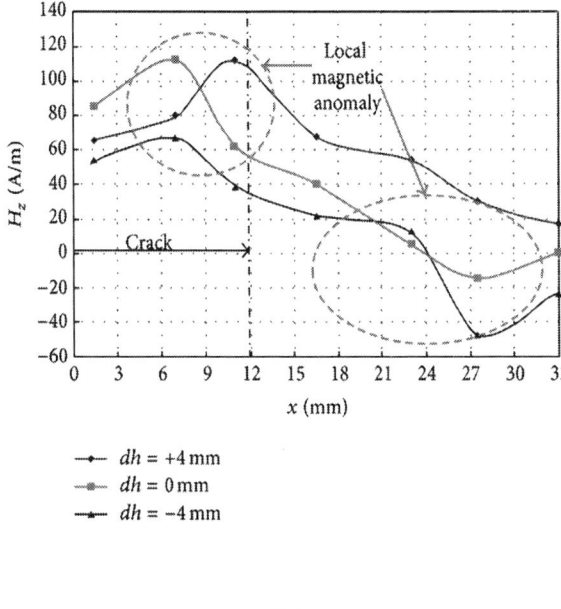

(b)

Figure 22: MMM symptoms of cracked blade after HCF test [3, 22].

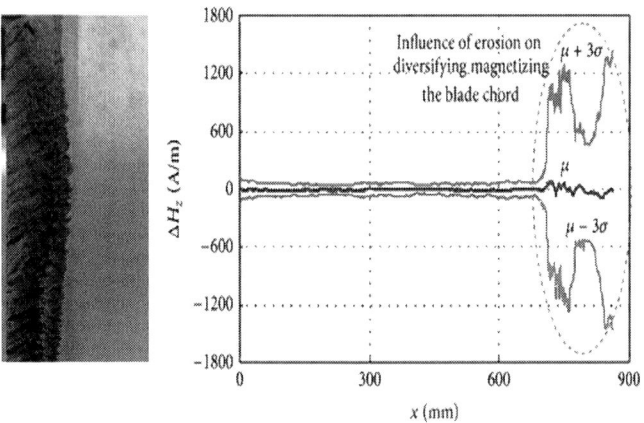

(a)

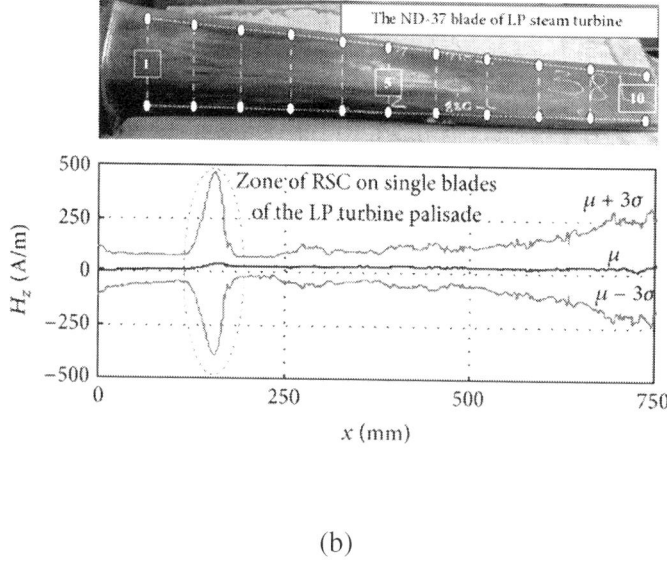

(b)

Figure 23: Early detection of the hidden fatigue risk (blade overload and diversifying the income of erosion) on the example of the ND37 blades of the steam turbine [3].

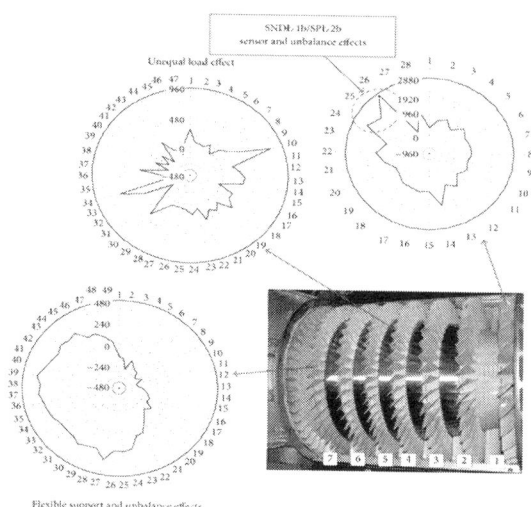

Figure 24: Detection of stress prehistory (irreversible process of stress magnetization) and identification of blade fatigue risk [3, 22].

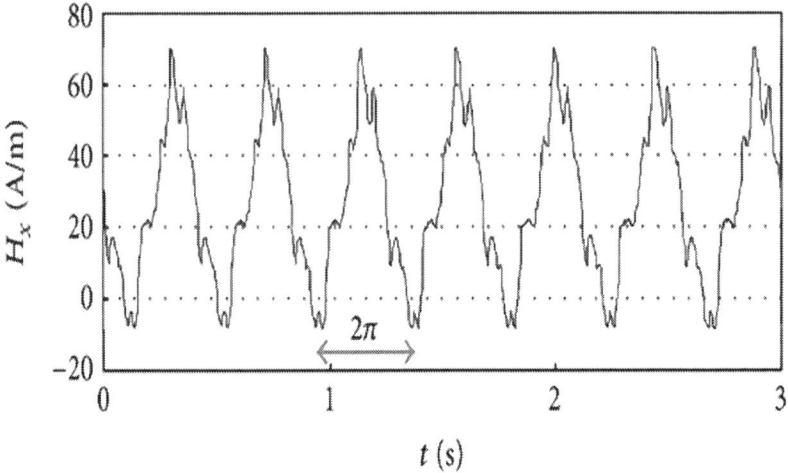

(a)

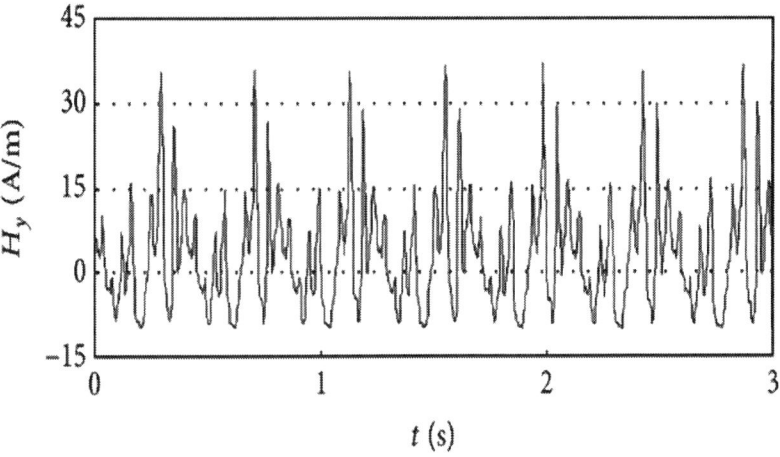

(b)

High Sensitive Methods for Health Monitoring of Compressor Blades...

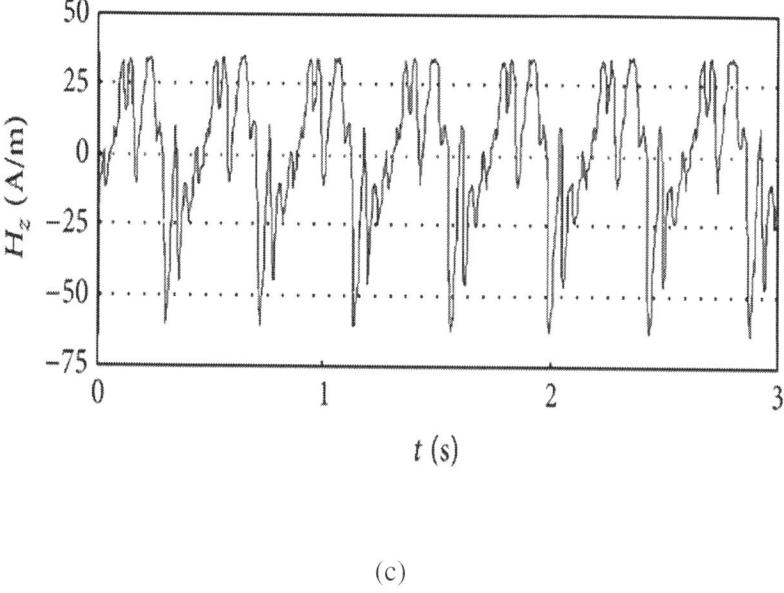

(c)

Figure 25: Components of the magnetic field measured on the suface of compressor casing (H_x: along the pivot of the engine, H_y: tangensial to the casing, and H_z: normal to the surface) [3].

The Experimental Modal Analysis Method

Experiments have been performed in five stages in which
- the measurement method has been verified,
- blade modal properties have been identified,
- blade cracking symptoms have been identified,
- early symptoms of fatigue have been identified,
- new diagnostic symptoms have been verified for titanium blade.

Identification of the Modal Properties

It was proven that PZT exciter and Polytec scanning vibrometer could be used for the modal identification of compressor blades. Measuring collected data is a verified knowledge used for tuning the FEM model up (Figure 26).

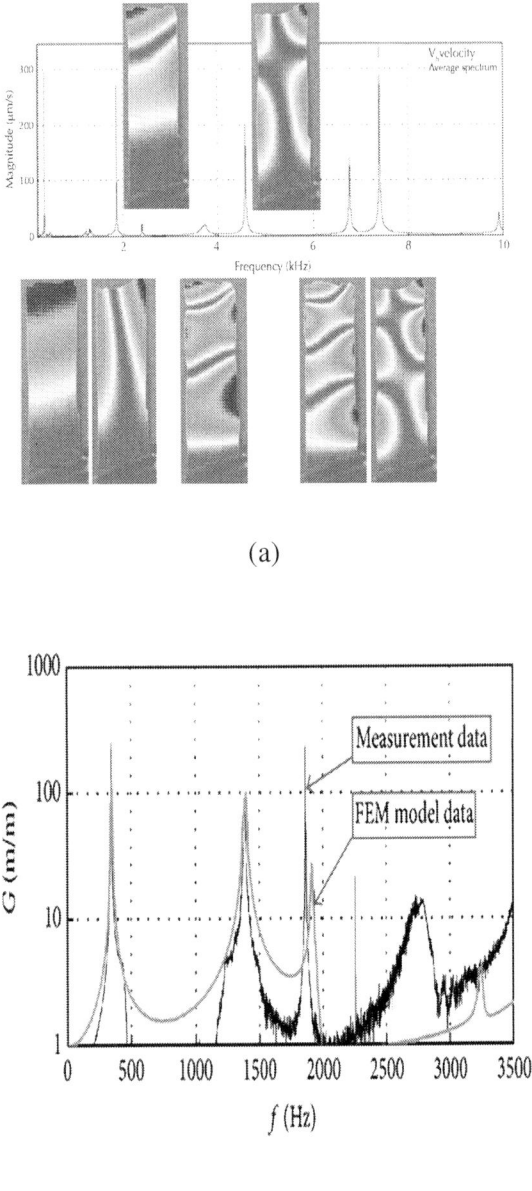

Figure 26: Identification of blade modal properties using (a) PZT exciter and laser scanning vibrometer (906 test points); (b) experimental data and FEM model of the steel blade (before the FEM model is tuning) [3, 22].

It has been also proven that used simple measurement technique (MTI laser head and sine test) guarantees reliable modal results when vibration amplitude is higher than 2 μm. Reliable resonance curve shape during sine test has been obtained for force frequency: 2.5 Hz/min for 1st flexible mode (1F, $Q_s > 350$) and 1.0 Hz/min for 1st torsion mode (1T, $Q_s > 1000$). Such a stand gives an ability to make precise measurements with an exact test profile and frequency step. The measurement system gives almost laboratory accuracy. That's why it let [23]:

- precise identification of blade modal properties in measured frequency range;
- metrological factors influence analysis on recorded resonance characteristics;
- modal parameters trends analysis be observed during fatigue tests.

Here in after of point 5, results will be presented from sine test. To analyze data we can use operator transmittance described by the following relation

$$G(\omega) = \frac{Y(\omega)}{X(\omega)} \left[\frac{m}{m} \right] \qquad (17)$$

with $X(\omega)$ being the magnitude of the exciter head displacement; $Y(\omega)$ being the magnitude of the blade point displacement; ω being angular frequency of cyclic load.

Modal Properties of a Defect-Free Blade (Noncracked)

In the case of a defect-free blade (health) resonance characteristics of particular modes were gained, ones that could be well described with a model of a single-degree-of-freedom linear system (SDOF)—of mass m suspended on a spring with spring rate K and viscous damping C [17–19, 23]. For sine test SDOF model describes the following relation:

$$m\frac{d^2y(t)}{dt^2} + C\frac{dy(t)}{dt} + Ky(t) = F(t),$$

$$F(t) = A(\omega)\sin(\omega t),$$

$$y(t) = B(\omega)\sin(\omega t + \varphi(\omega)). \tag{18}$$

Characteristics of subsequent modes remain continuous under resonance conditions and exhibit good symmetry around the resonance frequency (within the bandwidth of 3 dB) (Figure 27) (left side). The blade displacement at the measuring point can be described as follows

- vibration amplitude

$$b(\omega) = \frac{y_{st}}{\sqrt{\left[1 - (\omega/\omega_o)^2\right]^2 + ((\delta/\pi)(\omega/\omega_o))^2}}, \tag{19}$$

- (ii) vibration phase angle

$$\varphi(\omega) = \arctan\left(\frac{(\delta/\pi)(\omega/\omega_o)}{1 - (\omega/\omega_o)^2}\right), \tag{20}$$

where ω_0 is a free vibration frequency, δ is a logarithmic damping decrement

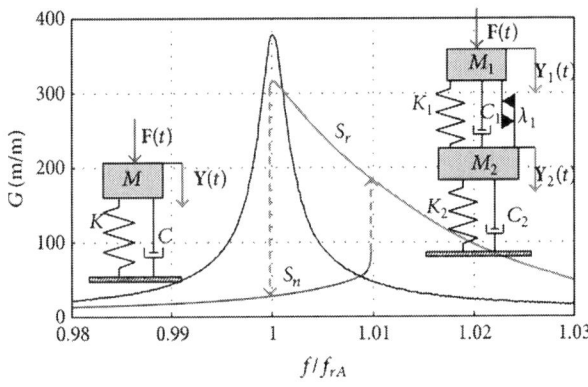

Figure 27: The effect of a crack on the 1st mode characteristics shape (steel blade, $a = 19.62$ m/s^2, f_{rA}: frequency of amplitude resonance) [23].

Diagnostic Symptoms of a Cracked Blade

When analyzing resonance curve shape we can observe how different it is for cracked blade. The blade has all nonlinear properties [17, 21, 23] which describe a nonlinear 2DOF model (Figure 27) (right side).

Close to the resonance frequency it is possible to observe two branches of characteristics: resonance attractor, S_n (red line) and nonresonance attractor, S_r (blue line) and jumps between them. The shape of a crackedblade's resonant curve is affected by the blade's material and conditions existing on the edge of the crack gap (weakening or hardening, friction). The characteristic curve is sloped to the left (towards lower frequencies) for the crack with material weakening. On the other hand, for the gap with material hardening, the curve is sloped to the right (towards higher frequencies) (Figure 28). The knowledge of resonant curve inclination is essential for correct interpretation of measurements, including correct identification of the resonant and nonresonant branches. During one-sided test we observe "asymmetry" resonance curve with seeming quality factor being decreased. Resonance frequency and characteristics are functions of a blade amplitude. They were not asymmetry symptom for the following.

- Small loads that do not develop an open crack: asymmetry is growing with a load increase.
- A notch on a blade, which was used as a simplified crack model (no friction at a notch hole): no friction in notch modeled blade is a source of other differences in modal properties, Table 4, and fatigue (JCF phenomena).

Table 4: Blade with 11mm length damage (starting from TE) placed 20mm from lock [23,25]

Blade	Frequency change (Hz)		
	1st mode	2nd mode	3rd mode
Cracked	−12	+7	−27
Notched (no friction)	−13	−5	−80
Difference (%)	**−0.28**	**−0.86**	**−2.73**

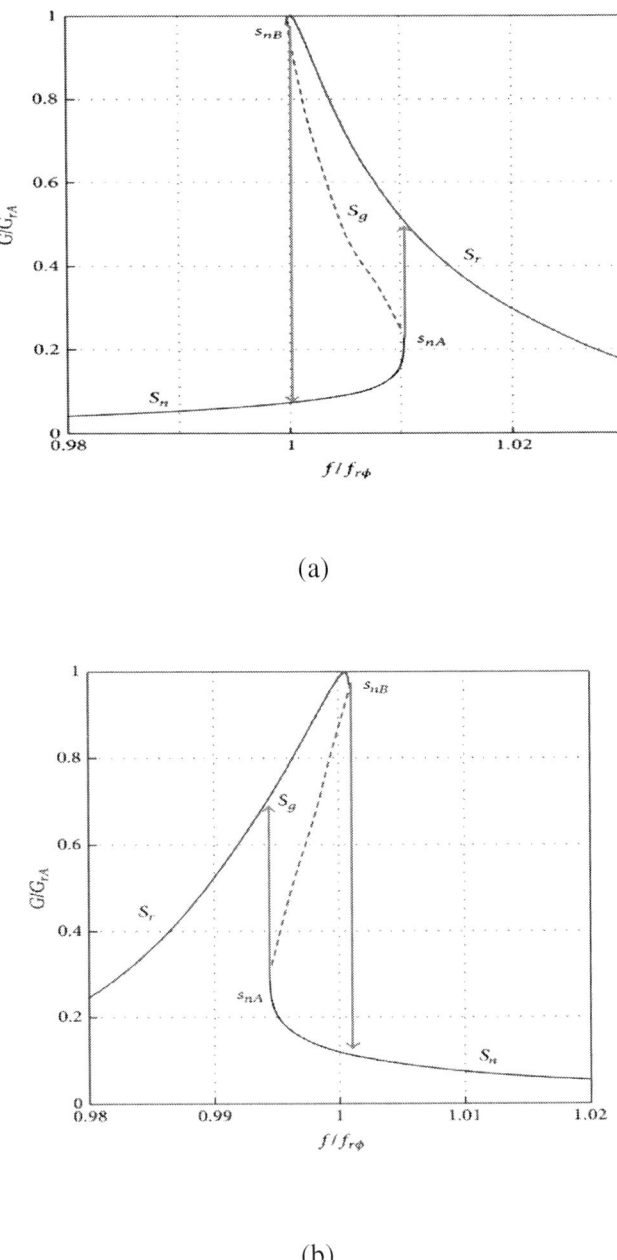

Figure 28: Shape of resonance characteristic for cracked blade with: (a) material weakening on the crack tip; (b) material hardening on the crack tip [3, 23].

The obtained characteristics of the cracked blade cannot be described with an SDOF linear model. The blade crack forms a two-degrees-of-freedom (2DOF) nonlinear system for any form of blade vibration. The equivalent linear equation that satisfies the nonlinear equation with accuracy ε takes the following form:

$$\frac{d^2 y}{dt^2} + 2h_\varepsilon(b)\frac{dy}{dt} + \alpha_\varepsilon^2(b)\, y = \varepsilon p \cos(\omega t), \tag{21}$$

where ε is small parameter, p is amplitude of the exciting force, b the steady-state vibration amplitude, $\alpha_\varepsilon(b)$ the equivalent natural (free-vibration) frequency, and $h_\varepsilon(b)$ the equivalent elementary damping coefficient. The measured and analyzed parameters of the blade are described with the following relationships:

- vibration amplitude

$$b(\omega) = \frac{\varepsilon p}{\sqrt{\left(\alpha_\varepsilon^2(b) - \omega^2\right)^2 + 4h_\varepsilon^2(b)\omega^2}}, \tag{22}$$

- resonance frequency

$$\omega = \sqrt{\left(\alpha_\varepsilon^2(b) - 2h_\varepsilon^2(b)\right) \pm d},$$

$$d = \sqrt{4h_\varepsilon^2(b)\left(h_\varepsilon^2(b) - \alpha_\varepsilon^2(b)\right) + \left(\frac{\varepsilon p}{b}\right)^2}, \tag{23}$$

- vibration phase angle

$$\varphi(\omega) = \arctan\left[\frac{-2h_\varepsilon(b)\omega}{\alpha_\varepsilon^2(b) - \omega^2}\right]. \tag{24}$$

Early Fatigue Identification

The LCF and HCF data analysis showed that blade modal properties could be used to observe the material strengthening phase [23].

Increase in the 1st mode resonance frequency of approximately 0.4% and reducing material damping are symptoms of the initial resonance system quality factor growth (correlation with structural and magnetic anisotropy) (Figure 29). This phase can be described with linear SDOF model.

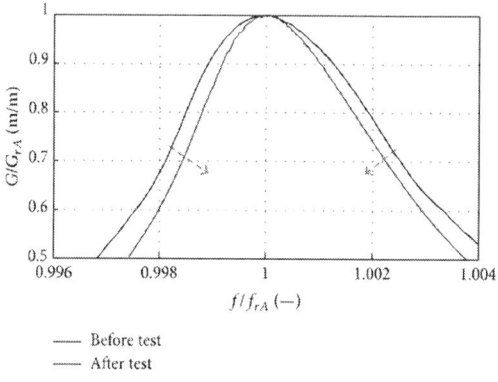

Figure 29: Changes in modal parameters during material strengthening phase [3, 23].

The growing asymmetry of the resonance curve was observed only in the final fatigue phase (Figure 30); it preceded the 1st mode frequency decrease.

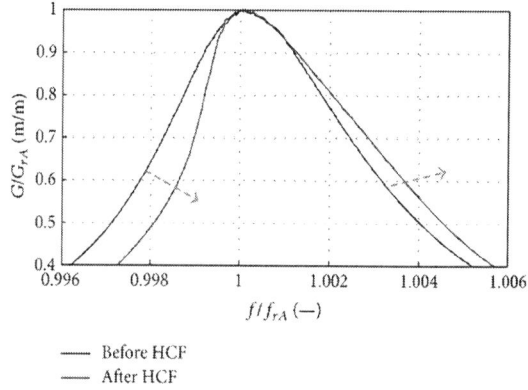

Figure 30: Changes in modal parameters during material weakening phase ($\delta f_{rA} = -0.5$ Hz) [3, 23].

JCF Phenomena

Influence of the cracked blade's resonant curve discontinuity on the propagation rate was tested for blades made from titanium alloy. It was found that in the case of constant frequency input (HCF tests without fine tuning to current resonant frequency), characteristic curve sloping to right and resonant curve discontinuity helps stopping the crack propagation.

The speed rate of its development was conditioned by the load history of a blade. The asymmetry is a symptom of the material weakening phase [3, 23]. The speed rate of the resonance curve asymmetry development, from the very first symptom of an open crack, is determined by the blade loading history.

Discontinuity of the resonant curve (blade pulse input discharge and load even for constant external load) is a source of very fast crack propagation during frequency transient phase—the phenomenon is called Jump Cycle Fatigue (JCF) (Figure 31).

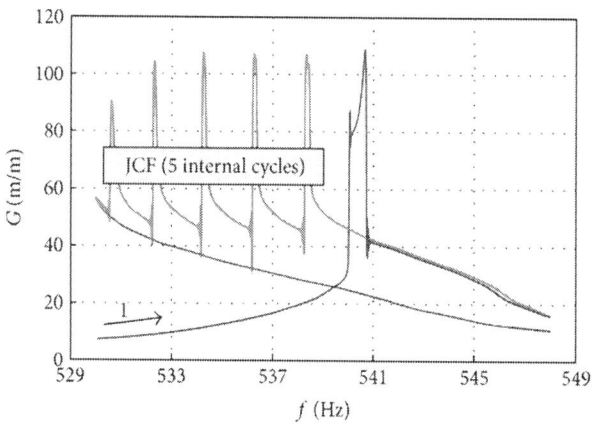

Figure 31: The JCF phenomenon for a cracking blade with Ti5.8Al-3.7Mo (sine sweep 4 Hz/min with constant external load) [3, 23].

The JCF is a reason for the serial material tearing during the decrease in excitations frequency, observed in the unstable phase of cracking. Those observations are fundamental for the prognosis of crack propagation velocity and determination safe prognosis horizon for blade operation and fatigue reverse engineering—correct interpretation of fracture structure (answer on the question "How many load cycles took place during crack propagation?"). Arrest lines of fatigue strap map only a number of cycles for internal loads. Their values could be bigger several times than a number of cycles for external loads, which result from a flight mission profile.

The Tip Timing Method

The object under scrutiny has been the 1st stage compressor blade (28 blades made out of the 18H2N4WA steel, each 100 mm long, chord 37 mm, twisted by the angle of 38°). Frequencies of three subsequent modes of blade vibration were as follows (average values): 350 Hz and 1380 Hz (bending vibration), 1890 Hz (torsional vibration).

Synchronous Resonance

During examination with a strain gauge no evident symptoms of interrelationships between the disk and blade vibrations were observed— compressor stages are of compact design. However, it was observed that, within the take-off range of the SO-3 engine operation (n = 15600 rpm), synchronisation of blade vibration with forces from the 2nd harmonic of the rotational speed (f_{1mode} = 520 Hz) may occur (Figure 32(a)). Such phenomena observed, for example, after some foreign object (bird, ice) has been deposited on the stator blade-ring, induce blade vibration up to some dangerous level where the material yield point is reached and exceeded, and quick initiation and propagation of the LCF and HCF cracks occur. Under such conditions of blade operation, time of safe operation of any turbojet engine may be much shorter than one flight/mission of an aircraft.

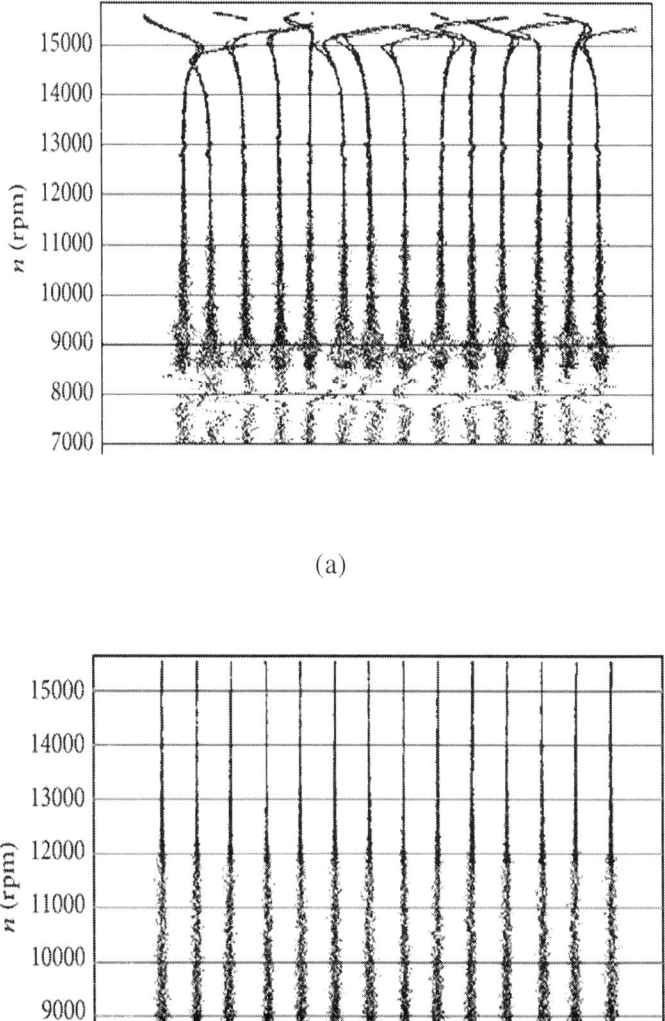

Figure 32: How foreign-matter depositions may affect the level of stress in the SO-3's 1st stage compressor blades [55]: (a) effect of foreign matter in compressor inlet; (b) model amplitude-phase spectra of the 1st compressor blade.

Asynchronous Resonance

The work of an engine in the compressor limits (during acceleration and deceleration of the rotation speed) exaggerated shading in the intake whether exaggerated mistakes of the rotor alignment are creating condition for the asynchronous resonance of compressor blades and HCF problems (Figure 33). They are not only endangered blades of the compressor but also other sub-assemblies of the engine, for example, bearing, gears, and shafts. The endurance risk of subassemblies mentioned above can be reduced through the engine user. For that purpose an aperiodic component TOA(k) is being used (phase portrait of the rotation speed) [61–63].

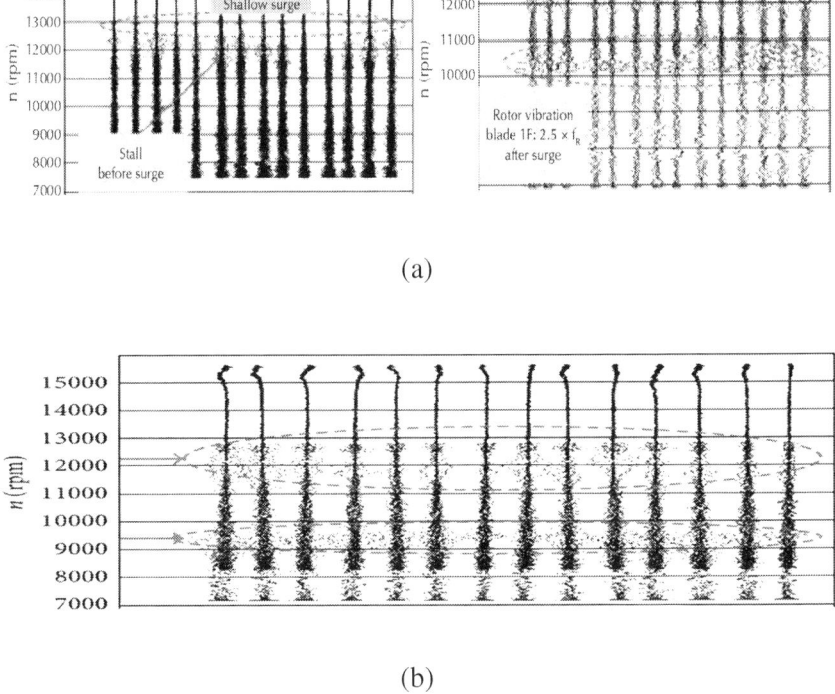

Figure 33: The asynchronous resonance of the compressor blades [55, 61, 63]: (a) before, during, and after the surge; (b) during identification of the surge limit (p_3 signal disconnected from FCU).

The Blade Cracking

After an analysis of destructive testing results (controlled propagation of blade cracking under normal conditions of operating the SO-3 engine) it was found that [55]

- during the blade cracking initiation (no open crack visible on the blade surface) only change in the B factor of dynamic increment of blade vibration frequency is seen (Figure 34(a)); frequency of the blade's free vibration $f_B(0)$ is constant

$$f_B(n) \cong \sqrt{f_B(n=0)^2 + Bn^2}, \qquad (25)$$

- the occurrence of a blade crack decreases in the range of excitations from the rotational-speed II harmonic by 1000 rpm ($\Delta f = 16.6$ Hz) (Figure 34(b)). At the moment, frequency (the 1st mode) of the blade's free vibration changed by less than 3 Hz;
- when the crack reaches about 30% of the blade profile, evident reduction in frequency of free vibration and decrease in the range of excitations from the rotational-speed III harmonic ($n \cong 8000$ rpm) were observed;
- just before the blade break-off (about 65% of profile for the crack from the leading edge, 95% of profile for the crack from the back of the blade), an evident effect of stiffening due to centrifugal forces was observed (Figure 34(c)). Changes in the dynamic scale inflicted by the broken blade are comparable with those in other dynamic scales (the influence of the engine's rotational speed).

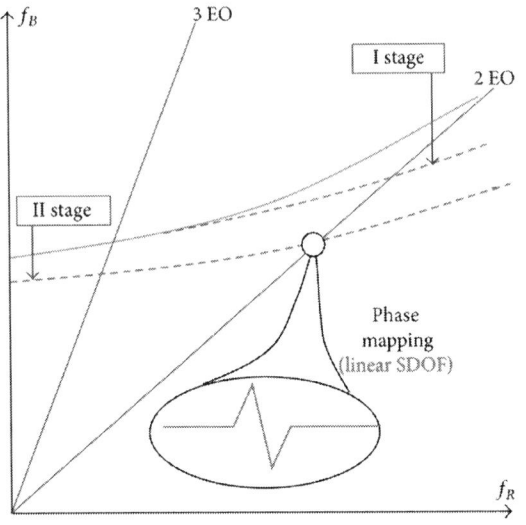

(a)

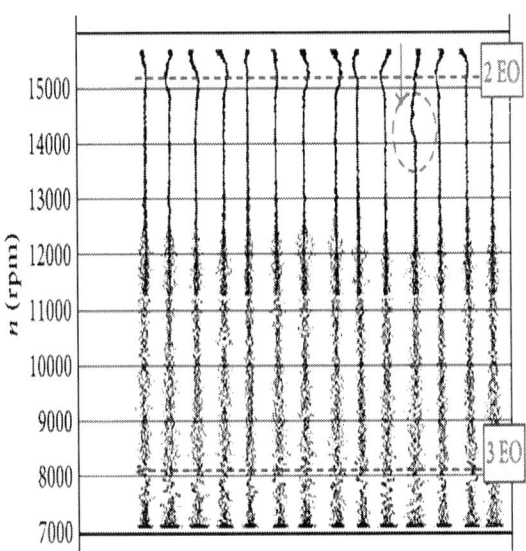

(b)

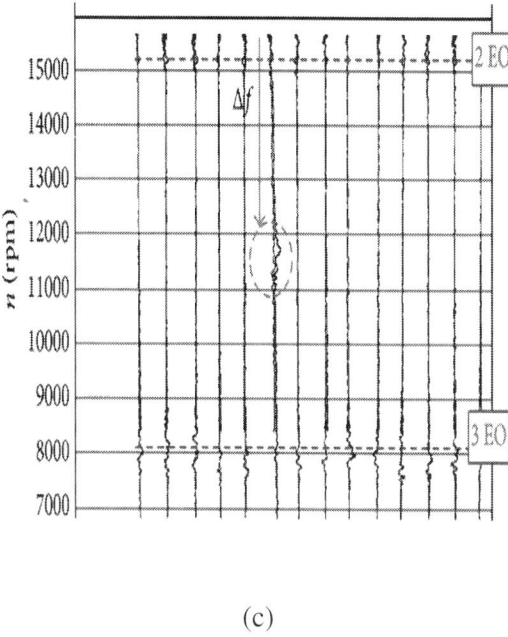

(c)

Figure 34: The effect of blade cracking [55]: (a) blade frequency plotted in the Campbell diagram; (b) the first stage of blade cracking, changes only B; (c) final stage of blade cracking, 5 minutes before break (signal after low-pass filtering).

It has been proven that the TTM gives credible prognosis for 50 engine work hours—"over 9.10^7 HCF and 100 LCF cycles, 1/8 TBO [3, 23, 62]." It has been also proven that TTM symptoms of the cracking are closely related to

- the strengthening phase: the quality factor of the resonance system increases together with the friction mode frequency;
- the weakening phase: growth in the resonance curve asymmetry and growth in nonlinearity.

CONCLUSIONS

- Described methods are mutually supplementing, which results in the synergy effect. The verified knowledge enables better modeling of continuum damage mechanisms and improving research method.

- New opportunities of the metal magnetic memory method, including diagnosing and identification of hidden risk of material fatigue (before the opened crack, measurements through the casing, and SHM application), have been proven.
- The high effectiveness of the experimental modal analysis method has been demonstrated on the basis of tests of more than 3000 compressor blades. A possibility of the automatic detection of the crack, the weakness (fatigue softening) and strengthening (strain hardening) of the blade, has been presented. The shape of resonance curve is diagnostic symptom.
- Active control of blade fatigue by the aeroengine user is possible. During 20 years of using the tip timing method in the Armed Forces of Poland, the following things have occurred.
    - The statistical mean time between fatigue break-offs of blades has been increased (eleven times for calendar-based data and seven times on the hourly basis). Since 1991 the fatigue crack of any compressor blade in the SO-3 engines has not been registered in spite of the existing fault in design.
    - The surge as a result of maladjustment of the fuel system and latent defects of subsystems has been eliminated (mainly fatigue problems results from maintenance).
    - Five SO-3 engines have been taken out of service due to excessive errors in shapes of the blades.
- Nonlinear properties of a crack blade are fundamental for the prognosis of the crack propagation rate and for the determining safe prognosis horizon. The modal symptoms of material damage are correlated with magnetic symptoms. Asymmetry of resonant curve has not been found on the blade with notched-a simple crack simulation model, often found in the literature.

ACKNOWLEDGMENTS

The study has been prepared under some research projects financially supported by the National Centre for Research and Development of Poland, the Ministry of Science and Higher Education of Poland, and the Ministry of National Defence of Poland. In examinations of compressor blades a staff of Institute of Fluid Flow Machinery Polish Academy of Science, aviation units, and the Military Aviation Works No. 1 gave the technical assistance.

REFERENCES

1. "More intelligent gas turbine engines," RTO-TR-AVT-128, NATO, 2009, http://www.cso.nato.int/pubs/rdp.asp?RDP=RTO-TR-AVT-128.
2. "Active control of engine dynamics," RTO-EN-020, AVT-083, NATO, November 2002, RTO AVT/VKI Special Course, 2001, Rhode-Saint-Genese, Belgium, http://www.cso.nato.int/Pubs/rdp.asp?RDP=RTO-EN-020.
3. M. Witoś, "Increasing the durability of turbine engines through active diagnostics and control," Research Works of AFIT, no. 29, 324 pages, 2011.
4. V. T. Vlasov and A. A. Dubov, Physical Bases of the Metal Magnetic Memory Method, ZAO "Tisso" Publishing House, Moscow, Russia, 2004.
5. ISO-24497-1. Non-destructive testing-Metal magnetic memory. Part 1: Vocabulary, 2007.
6. ISO-24497-2. Non-destructive testing-Metal magnetic memory. Part 2: General requirements, 2007.
7. ISO-24497-3. Non-destructive testing-Metal magnetic memory. Part 3: Inspection of welded joints, 2007.
8. A. Dubov and S. Kolokolnikov, "The metal magnetic memory method application for online monitoring of damage development in the steel pipes and welded joints specimens," Welding in the World, vol. 57, no. 1, pp. 123–136, 2013.
9. K. Yan, Z. D. Wang, B. Deng, and K. Shen, "Experimental research on metal magnetic memory method," Experimental Mechanics, vol. 52, no. 3, pp. 305–314, 2012.
10. M. Roskosz, "Metal magnetic memory testing of welding joints of ferritic and austenitic steels," inproceedings of the 40th International Conference and Exhibition, pp. 219–228, Pilsen, Czech, NDT for Safety/Defektoskopie, 2010.

11. X. Ding, J. Li, F. Li, and X. Pang, "Magnetic memory inspection of high pressure manifoolds," inProceedings of the 17th World Conference on Non-Destructive Testing, Shanghai, China, October 2008.
12. Q. Liu, J. Lin, M. Chen et al., "A study of inspecting the stress on downhole metal casing in oilfields with magnetic memory method," in Proceedings of the 17th World Conference on Nondestructive Testing, Shanghai, China, October 2008.
13. X. Hai Yan, X. M. Qiang, Y. Zhijun, and Z. Lihong, "Stress state Analysis of failure blade with MMM method," http://www.paper.edu.cn/.
14. M. Iwaniec, M. Witoś, M. Roskosz, and S. Gontarz, "Diagnosis of bearer structures of high voltage lines using magneto-mechanical effects," in Proceedings of the 4th Scientific and Technological Conference Diagnostics of Material and Industrial Components DMiUT, Gdansk, Poland, May 2012.
15. M. Witoś, "The reference signal of geomagnetic field for MMM expert systems," Key Engineering Materials, vol. 518, pp. 384–395, 2012.
16. M. Witoś and M. Zieja, "High sensitive methods for fatigue detection," Diagnostyka, vol. 3, no. 59, pp. 25–34, 2011, Journal of KONBiN, 2011, vol. 1, no. 17, pp. 307–326.
17. N. M. M. Maia and J. M. M. Silva, Theoretical and Experimental Modal Analysis, Research Studies Press LTD, Taunton, Mass, USA, 1997.
18. D. J. Ewins, Modal Testing: Theory, Practice and Application, Research Studies Press LTD, Baldock, UK, 2nd edition, 2000.
19. W. Heylen, S. Lammens, and P. Sas, Modal Analysis Theory and Testing, KUL Press, Leuven, Belgium, 1997.
20. B. J. Schwarz and M. H. Richardson, Experimental Modal Analysis, CSI Reliability Week, Orlando, Fla, USA, 1999.
21. L. A. Ostrovsky and P. A. Johnson, "Dynamic nonlinear elasticity in geomaterials," Revista Del Nuovo Cimento, vol. 24, no. 7, pp. 1–46, 2001.
22. M. Witoś and M. Stefaniuk, "Compressor blade fatigue diagnostic and modeling with the use of modal analysis," Fatigue of Aircraft Structure, vol. 1, no. 3, pp. 112–133, 2010.

23. M. Witoś, "On the modal analysis of a cracking compressor blade," Research Works of AFIT, no. 23, pp. 21–36, 2008.
24. "Report on aircraft PP-VNN," engine failure, 2000, http://aviation-safety.net/database/.
25. J. F. Brouckaert, Tip Timing and Tip Clearance Problems in Turbomachinary, Lecture Series 3-2007, VKI Belgium, 2007.
26. Agilis Non-Intrusive Stress Measurement System, Fundamentals, http://www.agilismeasurementsystems.com/.
27. B. W. Ayes, S. Arnold, C. Vining, and R. Howard, "Application of generation 4 non-contact stress measurement system on HCF demonstrator engines," in Proceedings of the 10th National Turbine Engine High Cycle Fatigue (HCF) Conference, Dayton, Ohio, USA, 2005.
28. F.-J. Duan, Z.-Q. Fang, Y.-Y. Sun, and S.-H. Ye, "Real-time vibration measurement based on tip-timing for rotating blades," Opto-Electronic Engineering, vol. 30, no. 1, pp. 29–31, 2005.
29. A. von Flotow and M. J. Drumm, Engine Sensing Technology Hardware and Software to Monitor Engine Rotor Dynamics Using Blade Time-of-Arrival and Tip Clearance, Hood River, Ore, USA, 2002.
30. R. Przysowa and J. Spychala, "Health monitoring of turbomachinery based on blade tip-timing and tip-clearence," RTO-MP-AVT-157 Paper 14, NATO, 2008.
31. R. Washburn, "Amplitude and phase variations associated with low order resonance responses subjected to time varying excitation sources," in Proceedings of the 9th National Turbine Engine High Cycle Fatigue Conference, Dayton, Ohio, USA, 2004.
32. M. Witoś, "Turbine engine health/maintenance status monitoring with use of tip timing method," in Proceedings of the 4th Workshop on Structural Health Monitoring, T. Uhl, W. Ostachowicz, and J. Holnicki-Szulc, Eds., pp. 157–164, DEStech Publication, 2008.
33. M. Zielinski and G. Ziller, "Noncontact crack detection on compressor rotor blades to prevent further damage after HC-failure," RTO MP-AVT-121 paper 19, NATO, 2005.
34. W. Campbell, "Elastic-fluid turbine rotor and method of avoiding tangential bucket vibration therein," Patent US 1, 502, 904, 1924.

35. http://www.evi-gti.com/.
36. http://www.piwg.org/.
37. G. H. Hardigg and P. A. Swarthmore, "Apparatus for measuring rotor blade vibration," Patent US 2, 575, 710, Westinghouse Electric Corporation, 1951.
38. H. Shapiro, "Vibration detector and measuring instrument," Patent US 3058339, Curtiss-Wright Corporation, Propulsion Products Division, 1962.
39. I. E. Zablotsky, J. A. Korostelev, A. Y. Lebedev, L. B. Sviblov, and E. M. Tolchinsky, "Vibrator indicator for turboengine rotor blading," Patent US 3, 467, 358, 1969.
40. J. Smejkal, M. Jindra, and Z. Brezina, "Apparatus for switching pulses in measuring the vibration of rotating parts during operation of a machine," Patent US 3, 597, 963, Skoda Czechoslovakia, 1971.
41. J. Naegeli and A. Maurer, "Method and apparatus for monitoring the state of oscillation of the blades of a rotor," Patent US 4, 153, 388, Sulzer Brothers Ltd, 1979.
42. V. E. H. Ellis, "Vibration monitoring in rotary machines," Patent US 4, 593, 566, Rolls Royce Ltd, 1986.
43. G. I. Marron and W. B. Rethage, "Blade pitch measurement apparatus and method," Patent US 4, 827, 435, Westinghouse Electric Corporation, 1989.
44. F. S. McKendree and P. F. Rozelle, "Nonsynchronous turbine blade vibration monitoring system,"Patent US 4, 887, 468, Westinghouse Electric Corporation, 1989.
45. R. P. Kending, R. A. Lucheta, and F. S. McKendree, "Turbine blade fatigue monitor," Patent US 4, 955, 269, Westinghouse Electric Corporation, 1990.
46. R. Kudelski and R. Szczepanik, "System for signalling the fact of exceeding admissible amplitude of vibrations by vanes of a fluid-flow machine," Patent 184530 (PL), Air Force Institute of Technology, 1991.
47. M. Twerdochlib, P. F. Rozelle, and S. Sarasas, "Apparatus and method for removing common mode vibration data from digital turbine blade vibration data," Patent US 5, 148, 711, Westinghouse Power Corporation, 1992.

48. M. Witoś, A. Gawin, and A. Szczepankowski, "Sposob diagnozowania technicznego wirujacych lopatek maszyny wirnikowej oraz uklad do diagnozowania technicznego wirujacych lopatek maszyny wirnikowej," Patent PL 169219 (B1), Instytut Techniczny Wojsk Lotniczych, 1996.
49. R. P. Kendig, R. A. Lucheta, and F. S. McKendree, "Turbine blade fatigue monitor," PatentKR960014013 (B1), Westinghouse Electric Corporation, 1996.
50. M. Yukio and E. Masanori, "Vibration measuring apparatus for rotor blade," Patent JP3038382 (B2), NATL Aerospace Lab, 2000.
51. V. A. Mednikov and V. V. Shchegolev, "Method for determining amplitude of oscillations of turbomachine blade," Patent RU2207524 (C1), 2003.
52. I. Hideyasu, N. Kenichi, and M. Shinya, "Moving blade failure diagnosing method of gas turbine and device therefor," Patent JP3663609 (B2), Ishikawajima Harima Heavy Ind, 2005.
53. V. V. Shchegolev, "Method for metering vibration amplitudes of turbomachine blades," PatentRU2244272 (C1), 2005.
54. M. Zielinski and G. Ziller, "Method and device for detecting cracks in compressor blades," PatentWO2010054644 (A2), MTU Aero Engines Gmbh, 2010.
55. M. Witoś, Diagnosing of technical condition of turbine engine compressor blades using non-contact vibration measuring method [Ph.D. thesis], ITWL, Warszawa, Poland, 1994.
56. A. A. Shaniavski, Modeling of Fatigue Cracking of Metals. Synergetics For Aviation, Publishing House of Scientific and Technical Literature 'Monography', Ufa, Russia, 2007.
57. A. A. Shaniavski, Tolerance Fatigue of Aircraft Components. Synergetics in Engineering Applications, Publishing House of Scientific and Technical Literature 'Monography', Ufa, Russia, 2003.
58. Y. Murakami, T. Nomoto, and T. Ueda, "Factors influencing the mechanism of superlong fatigue failure in steels," Fatigue and Fracture of Engineering Materials and Structures, vol. 22, no. 7, pp. 581–590, 1999.View at Publisher · View at Google Scholar · View at Scopus

59. Y. Murakami, M. Takada, and T. Toriyama, "Super-long life tension-compression fatigue properties of quenched and tempered 0.46% carbon steel," International Journal of Fatigue, vol. 20, no. 9, pp. 661–667, 1998. View at Scopus
60. T. Sakai, "Review and prospects for current studies on very high cycle fatigue of metal materials for machine structural use," Journal of Solid Mechanics and Materials Engineering, vol. 3, no. 3, pp. 425–439, 2009.
61. A. Szczepankowski, Diagnosing of Technical Condition of Turbine Engine Using Rotational Speed Phase-Mapping Method [Ph.D. thesis], ITWL, Warszawa, Poland, 1999.
62. R. Szczepanik and M. Witoś, "Aeroengine condition monitoring system based on non-interference discrete-phase compressor blade vibration measuring method," RTO-MP-051, NATO, 2001, http://ftp.rta.nato.int/public//PubFullText/RTO/MP/RTO-MP-051///MP-051-PSP-13.pdf.
63. M. Kowalski, "Phase mapping in the diagnosing of a turbojet engine," Journal of Theoretical and Applied Mechanics, vol. 50, no. 4, pp. 913–921, 2012.
64. S. Kocanda, Zmeczeniowe Pekanie Metali, WNT, Warszawa, Poland, 1985.
65. M. A. Almojil, Deformation and recrystal-lisation in low carbon steels [Ph.D. thesis], Manchester Materials Science Centre, 2010.
66. V. Novikov, Grain Growth and Control of Microstructure and Texture in Polycrystalline Materials, CRC Press, Boca Raton, Fla, USA, 1996.
67. A. H. Cottrell, The Mechanical Properties of Matter, John Wiley and Sons, New York, NY, USA, 1964.
68. F. R. N. Nabarro, "Dislocations in a simple cubic lattice," Proceedings of the Physical Society, vol. 59, no. 2, article 309, pp. 256–272, 1947. View at Publisher · View at Google Scholar · View at Scopus
69. N. A. Makhutov and M. M. Gadenin, "Nonlinear deformation and fracture mechanics for engineering approaches in design of structure," Mechanical Engineering, Energy Systems and Sustainable Development, vol. 1, http://www.eolss.net/.

70. W. Cui, "A state-of-the-art review on fatigue life prediction methods for metal structures," Journal of Marine Science and Technology, vol. 7, no. 1, pp. 43–56, 2002. View at Publisher · View at Google Scholar · View at Scopus
71. G. Socha, "Experimental investigations of fatigue cracks nucleation, growth and coalescence in structural steel," International Journal of Fatigue, vol. 25, no. 2, pp. 139–147, 2003. View at Publisher ·View at Google Scholar · View at Scopus
72. F. B. Pickering, Physical Metallurgy and the Design of the Steels, Applied Science Publishers, London, UK, 1978.
73. S. M. Thompson, The magnetic properties of plastically deformed steels [Durham theses], Durham University, 1991, http://etheses.dur.ac.uk/3600/.
74. T.-K. Lee, J. W. Morris Jr., S. Lee, and J. Clarke, "Detection of fatigue damage prior crack initiation with scanning SQUID microscopy," in Review of Progress in Quantitative Nondestructive, vol. 25, pp. 1378–1385, August 2005. View at Publisher · View at Google Scholar · View at Scopus
75. R. Newnham, Properties of Materials. Anisotropy, Symmetry, Structure, Oxford University Press, Oxford, UK, 2005.
76. L. Vandenbossche, Magnetic Hysteretic Characterization of Ferromagnetic Materials with Objectives towards Non-Destructive Evaluation of Material Degradation [Ph.D. thesis], Universiteit Gent, 2009.
77. J. A. Ewing, Magnetic Induction in Iron and Other Metals, "THE ELECTRICIAN" Printing and Publishing, London, UK, 1900.
78. K. Yaegashi, "Dependence of magnetic susceptibility on dislocation density in tensile deformed iron and Mn-steel," ISIJ International, vol. 47, no. 2, pp. 327–332, 2007.
79. W. Liang, D. Fang, Y. Shen, and A. K. Soh, "Nonlinear magnetoelastic coupling effects in a soft ferromagnetic material with a crack," International Journal of Solids and Structures, vol. 39, no. 15, pp. 3997–4011, 2002. ··
80. C. W. Burrows, "Correlation of the magnetic and mechanical properties of steel," Scientific Papers of the Bureau of Standards, no. 272, 1916, http://www.archive.org/.

81. M. S. Blanter, I. S. Golovin, H. Neuhäuser, and H. R. Sinning, Internal Friction in Metallic Materials. A Handbook, Springer, Berlin, Germany, 2007.
82. M. F. Fischer, "Note on the effect of repeated stresses on the magnetic properties of steel," Bureau of Standards Journal of Research, vol. 1, no. 5, pp. 721–732, 1928.
83. I. M. Robertson, "Magneto-elastic behaviour of steels for naval applications," MRL Technical ReportMRL-TR-90-27, DSTO Materials Research Laboratory, 1991.
84. T. Yamasaki, S. Yamamoto, and M. Hirao, "Effect of applied stresses on magnetostriction of low carbon steel," NDT and E International, vol. 29, no. 5, pp. 263–268, 1996.
85. D. L. Atherton and D. C. Jiles, "Effects of stress on magnetization," NDT International, vol. 19, no. 1, pp. 15–19, 1986.
86. R. R. Birss and C. A. Faunce, "Stress-induced magnetization in small magnetic fields," Journal de Physique, vol. 32, supplement, no. 2-3, pp. 686–688, 1971.
87. E. W. Lee, "Magnetostriction and magneto-mechanical effects," Reports on Progress in Physics, vol. 18, pp. 184–229, 1955.
88. C. M. Smith and G. W. Sherman Jr., "A study of the magnetic qualities of stressed iron and steel,"Physical Review, vol. 4, pp. 267–273, 1914.

Chapter 4

A Review on Recent Contribution of Meshfree Methods to Structure and Fracture Mechanics Applications

S. D. Daxini[1] and J. M. Prajapati[2]

[1]Department of Mechanical Engineering, Babaria Institute of Technology, Vadodara, Gujarat 391240, India

[2]Department of Mechanical Engineering, M. S. University, Vadodara, Gujarat 390020, India

ABSTRACT

Meshfree methods are viewed as next generation computational techniques. With evident limitations of conventional grid based methods, like FEM, in dealing with problems of fracture mechanics, large deformation, and simulation of manufacturing processes, meshfree methods have gained much attention by researchers. A number of meshfree methods have been proposed till now for analyzing complex problems in various fields of engineering. Present work attempts to review recent developments and some earlier applications of well-known meshfree methods like EFG and MLPG to various types of structure mechanics and fracture mechanics applications like bending, buckling, free vibration analysis, sensitivity analysis and topology optimization, single and mixed mode crack problems, fatigue crack growth, and dynamic crack analysis and some typical applications like vibration of cracked structures, thermoelastic crack problems, and failure transition in impact problems. Due to complex nature of meshfree shape functions and evaluation of integrals in domain, meshless methods are computationally expensive as compared to conventional mesh based methods. Some improved versions of original meshfree methods and other techniques suggested by researchers to improve computational efficiency of meshfree methods are also reviewed here.

INTRODUCTION

Numerical simulation has proved to be a good alternative scientific investigation tool to expensive, time consuming, and sometimes dangerous experiments in complex engineering problems. Grid based numerical methods, like FEM, are widely used for analyzing various engineering problems. There are two fundamental approaches in grid based methods: Eulerian and Lagrangian grid. To strengthen the advantages of each approach and avoid their limitations, new combined approaches were also developed [1]. But grid based methods are not well suited to treat the problems of fracture mechanics with moving material discontinuity, large deformation problems where excessive mesh distortion takes place, and when simulation of some manufacturing process is to be studied.

By modifying the internal structure of gird based method, meshfree methods were developed which are expected to be more adaptive, versatile, and robust and can deal with problems where conventional methods are not suitable. The concept of the meshfree methods is to provide accurate and stable numerical solutions for integral equations or PDEs with all types of possible boundary conditions with a set of arbitrarily distributed nodes without defining mesh which connects these nodes [1]. Many meshfree methods have been developed till now. Earliest meshfree method was developed in 1977 by Lucy and Gingold and Monaghan as smoothed particle hydrodynamics (SPH), a meshfree particle method [2–4]. It was initially developed for modeling astrophysical phenomena but later widely used for applications of solid and fluid mechanics. Many corrected versions of SPH were proposed by researches to solve problems of instabilities and inconsistencies in original SPH model. Later, Nayroles et al., in 1992, were the first to use moving least square approximations in a Galerkin method to formulate the so-called diffuse element method (DEM). Based on the DEM, Belytschko et al., in 1994, advanced remarkably and proposed the element free Galerkin (EFG) method, which was the first meshfree method based on global weak form [5]. Atluri and Zhu, in 1998, had originated the meshless local petrov-Galerkin (MLPG) method, based on local weak form that requires only local background cells for the integration [6]. Liu and his coworkers proposed reproducing kernel particle method (RKPM) [7]. Subsequently, other meshfree methods were developed by researchers, like hp cloud method by Duarte and Odent 1996; point interpolation method (PIM) by Liu and Gu, 1999, Wang and Liu, 2000, 2001, 2002, and meshfree weak-strong form (MWS) by Liu and Gu 2002, 2003 [1]. Some of the important features of meshfree methods are as follows, which makes them superior [8]:(a) there is no mesh alignment sensitivity, and mesh used in background is for integration purpose;(b)node connectivity is not predefined by mesh;(c)no remeshing is required especially in case of large deformation and moving discontinuity problems;(d)shape functions of any desired order continuity can be constructed;(e)no postprocessing required for smooth derivatives of unknowns and their derivatives.

Procedural steps involved in developing solution using meshfree method are shown in Figure 1.

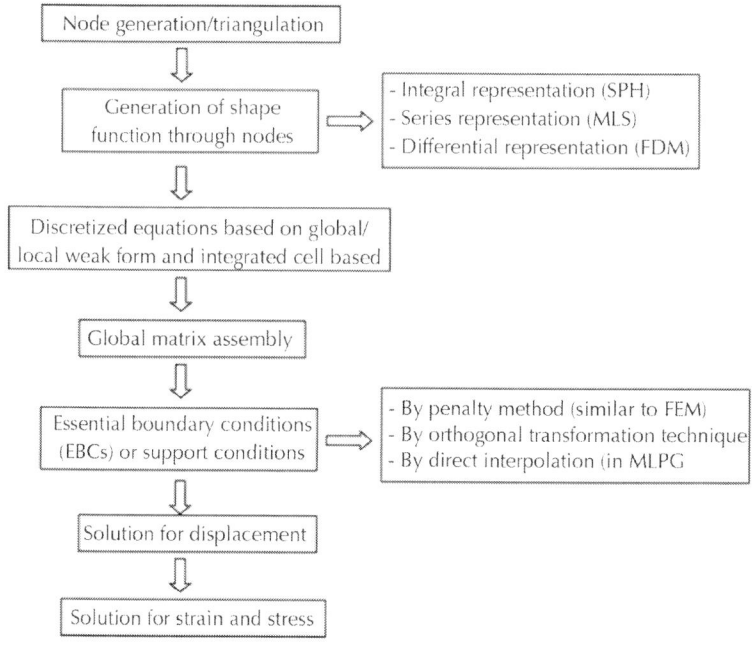

Figure 1: Procedural steps in meshfree methods.

Some review papers have also been presented in the area of development and applications of meshfree methods, earlier. Like, Belytschko et al. presented review on meshless approximation based on MLS, reproducing kernels (RK) and partition of unity methods (PUM) in 1996. The review included the following aspects: salient features of these methods, techniques to handle material and geometric discontinuity, implementation issues, like EBCs, coupling with finite elements, computational efficiency, convergence rate, and so forth, review of applications of plate and shell problems [9]. Li and Liu, in 2002, reviewed recent development of meshfree particle methods and their application in applied mechanics. Major approaches reviewed by them were SPH, meshfree Galerkin methods like DEM, EFG, MLPG, and hp cloud and some applications of molecular dynamics (MD) [10]. Nguyen et al., in 2008, presented a review on meshless methods and its computer implementation aspects with the aim of providing practical overview of meshless methods based on global weak form through a simple and well-structured MATLAB code including intrinsic and extrinsic enrichment, some boundary condition enforcement schemes and few one, and two dimensional numerical examples [11]. Liew et al., in 2011, presented their review on meshless methods for laminated and functionally graded plates and shells, wherein EFG and RKPM methods and their applications, including static and dynamic analysis, buckling, free vibration, and non-linear analysis, were in focus [12].

The present paper attempts to review recent and some earlier applications of some of the well-known meshfree methods like EFG and MLPG, without giving mathematical description, to structure and fracture mechanics problems. The outline of paper is as follows. Section 2 gives basic concepts of different aspects in meshfree methods like shape functions, weight functions, techniques for imposing essential boundary conditions and numerical integration, and so forth. Section 3 presents review of EFG and MLPG applications to various structure mechanics problems. Section 4 presents review of MMs applications to fracture mechanics problems. While review of some typical applications and techniques developed for improving computational efficiency of MMs is presented in Section 5 followed by conclusion and discussion of review in Section 6.

EFG AND MLPG

Meshfree methods are classified based on use of global or local weak form to derive system matrices. Accordingly, EFG method is based on global weak form, while MLPG method is based on local symmetric weak form (LSWF). In both these methods, approximation is based on moving least square (MLS) approximants. But in moving least square methods interpolants do not pass through data point as interpolation functions are not unity at nodes [5]. Hence, imposition of essential boundary conditions (EBCs) gets complicated in these methods. Present section provides basics of shape function construction, selection of weight functions, and techniques to impose essential boundary conditions and integration.

MLS Approximations

The basic idea of MLS approximation is based on construction of set of nodes in the problem domain and hence the method is element free [5].

The MLS approximant $u^h(x)$ of the function u(x) defined over the domain Ω is given by

$$u^h(x) = p^T(x)a(x), \quad \forall x \in \Omega, \tag{1}$$

where $p^T(x)$ are monomial basis functions of order m and a(x) are vector coefficients which are functions of space coordinates x, which can be determined at any point x by minimizing weighted discrete L_2 norm defined as follows,

$$J(x) = \sum_{i=1}^{n} w_i(x - x_i)\left[p^T(x_i)a(x) - u_i\right]^2, \tag{2}$$

where, n is the number of nodes in neighborhood of x for which weigh function $w_i(x-x_i)$ cannot be zero and u_i is the nodal value of u at $x=x_i$. On further solution for a(x) final expression for MLS approximants is given by

$$u^h(x) = \sum_{i=1}^{n} \Phi_i(x) u_i, \tag{3}$$

where $\phi_i(x)$ is called the shape function of MLS approximation. Detailed shape function construction can be referred from references [5, 13].

Selection of Weight Functions

Weight function selection is also an important parameter while developing the meshfree solution. It should be constructed in such a way that their value should decrease as the distance from x to x_i increase. Selected weight functions must be positive and the function and its derivative should be continuous up to required degree [5]. Some of the weight functions used are as the following.

Gaussian weight function:

$$w_i(x) = \frac{\exp\left[-(d_i/c_i)^{2k}\right] - \exp\left[-(r_i/c_i)^{2k}\right]}{1 - \exp\left[-(r_i/c_i)^{2k}\right]}, \quad 0 \le d_i \le r_i$$

$$= 0, \quad d_i \ge r_i, \qquad (4)$$

spline weight function:

$$w_i(x) = 1 - 6\left(\frac{d_i}{r_i}\right)^2 + 8\left(\frac{d_i}{r_i}\right)^3 - 3\left(\frac{d_i}{r_i}\right)^4, \quad 0 \le d_i \le r_i$$

$$= 0, \quad d_i \ge ri, \qquad (5)$$

where $d_i = |x - x_i|$ is the distance from node x_i to any point x, c_i is the constant controlling shape of the weight function, and r_i is the size of the support. There are several other types of weight functions used like conical weight function, cubic spline weight function, and so forth.

Imposing Essential Boundary Conditions

Because MLS shape functions used in EFG and MLPG do not satisfy Kronecker delta criterion, process of imposition of EBCs gets complicated than FEM. Number of techniques were developed for enforcing EBCs in the problem like Lagrange multiplier, penalty method, orthogonal transformation techniques, coupling with FEM, Nitsche's method, singular weighing functions, boundary collocation, and D'Alembert's Principle. Out of these techniques, penalty method can be easily implemented and do not increase much computational effort. Detailed description of these techniques can be referred from references [5, 14–19].

Integration Techniques

Several different techniques were suggested for numerical integration of Galerkin weak form. Gauss quadrature is most commonly employed technique to evaluate integrals in Galerkin weak form. Integration in meshfree method is based on background cells which are independent of nodes. Background cells serve important purpose of identifying nodes contributing to discrete L_2 norm at a quadrature point [8]. By minimizing the mismatch of shape function local support domain with integration cells, integration errors can be minimized and accuracy and convergence can be improved [20]. But because of some inherent drawbacks like complexity, requirement of higher order quadrature rules, specialized integration zone patterns, and so forth, direct nodal integration technique was proposed. But it led to oscillations in solution due to under-integration of weak form and vanishing shape functions at nodes. To alleviate this issue, a stabilized nodal integration technique was proposed by adding a residual of the equilibrium equation to the potential energy function which does not need background cell structure and results in completely meshless method [21, 22].

EFG AND MLPG: STRUCTURE MECHANICS

Meshless methods developed, in their original form, are not entirely "meshless" and each method falls in one of the following categories: methods based on global weak form requiring background cells for integration like EFG, methods based on local weak form requiring background cells locally like MLPG, and particle methods which require predefinition of particles for their volume or mass like SPH [8]. Following sections review applications of EFG and MLPG to structure and fracture mechanics problems.

Static and Dynamic Analysis

The element-free Galerkin (EFG) method was developed by Belytschko et al. based on the diffuse elements method (DEM) originated by Nayroles et al. In original form of EFG, moving least square (MLS) approximants were used to approximate field variables but due to lack of Kronecker delta property in MLS shape functions, essential boundary conditions (EBCs) cannot be imposed in straight forward way as in FEM and some special techniques are required. Lagrange multiplier technique was used for enforcing EBCs. Because EFG method is based on global weak form, it requires a mesh of background cells for integration in computing the system matrices. Proposed EFG method had advantages like high rate of convergence, no post processing for unknowns or their derivatives, and suitability to fracture mechanics problems [5]. Several techniques for enforcing EBCs in meshfree Galerkin method were proposed. The same authors proposed modified variational principle instead of Lagrange multiplier for imposing EBCs in EFG method [14]. A new technique for imposing EBCs in meshfree methods was proposed by Krongauz and Belytschko, wherein finite elements were used along the essential boundaries and shape functions of finite elements were combined with approximants used. High rate of convergence was observed with implementation of present technique [15]. Another boundary condition enforcement technique was proposed by Günther and Liu by a computationally efficient algorithm based on D'Alembert's principle that can be used for general constraints and fluid structure interface in meshless methods [16]. MLS shape functions used in EFG method are more complex than piecewise polynomial like shape functions used in FEM; hence Krysl and Belytschko presented a straightforward way to program the EFG shape function construction in a way which leads to both a simple interface to application code and to the implementation of EFG shape function itself [23]. Dolbow and Belytschko proposed EFG method implementation with its structured MATLAB code to benchmark structure problems in one-dimensional and two-dimensional applications. To enforce EBCs, few techniques were suggested like Lagrange multiplier, modified variational principles, and coupling with finite elements [24]. To solve three-dimensional elastic and elastoplastic problems, EFG method was proposed by Barry and Saigal with variable domain of influence approach. Singular weight functions were utilized in MLS shape functions allowing accurate and direct nodal imposition of EBCs. Several elastic and small strain elastoplastic problems were presented [25]. Tiago and Leitao applied EFG method to free vibration analysis of beams and plates. Shape functions were constructed by MLS approximation and kinematic boundary conditions were imposed by Lagrange multiplier technique in their work [26]. To improve computational efficiency of original EFG method, Zhang et al. presented improved EFG (IEFG) method by employing improved MLS (IMLS) shape functions for two-dimensional potential problems. MLS approximants yield precise solution but sometimes final algebra equations are ill-conditioned, which is undesirable. Improved MLS (IMLS) approximation was proposed by Liew et al. to alleviate the problem in boundary element method [6]. Proposed improved EFG method uses weighted orthogonal basis function for construction of MLS shape functions which avoids ill-conditioned algebra equations as in case of conventional MLS interpolation. IEFG method uses fewer nodes in entire domain than conventional EFG method and resulting in higher computation speed [27]. Proposed IEFG approach was extended for solving three-dimensional potential problems by same authors [28]. In a more recent development, for two dimensional elastoplasticity problems, complex variable moving least square approximation (CVMLS) and EFG based CVEFG were proposed by Peng et al. With CVMLS, it becomes possible to select fewer nodes in the meshless method than are required in the meshless method of the MLS approximation without loss of precision or in other words, CVMLS is computationally more efficient [29]. Presently, EFG method is one of the most popular meshfree methods, and applied to many structure and fracture problems, some of which are reviewed here in subsequent sections.

Another new meshfree computational method was developed and proposed by Atluri and Zhu, known as meshless Local Petrov Galerkin method (MLPG), based on local weak form and MLS approximants. EBCs are imposed by penalty method in MLPG. Selection of trial (shape) function and test function in MLPG is done from entirely different spaces and it is considered as a truly meshless method because all integrals can be easily evaluated over regular shaped domains and their boundaries. High convergence rate and accurate values of unknown variables and its derivatives were observed [13]. Elastostatic problems, like an infinite plate with circular and elliptical hole, were addressed by the same authors using MLPG method [30]. MLPG is a general concept; hence a comparison study of the efficiency and accuracy of a variety of meshless trial and test functions for different variants of MLPG was proposed by Atluri and Shen, wherein five types of trial functions and six types of test functions were explored and six different approaches, popularly known as MLPG1 to MLPG6, were presented. Numerical results for standard patch test, Laplace and Poisssion's equations, were compared for efficiency and computational cost and MLPG5 was found less expensive from computational view point as it employs local, nodal based test function over a local subdomain, a Heaviside step function [31]. Long and Atluri proposed MLPG for bending of thin (Kirchhoff) plates based on MLS approximants and local symmetric weak form (LSWF). Cubic, quartic, and quintic basis, as well as the quitic spline weight function was employed in the MLS computation, while EBCs were enforced by penalty method in proposed computation [32]. Raju and Phillips had shown MLPG application for Euler Bernoulli beam problems like cantilever beam and simply supported beam with various loading conditions by selecting simple weight functions as test functions and MLS approximation [33]. The same authors presented MLPG method for Euler-Bernoulli beams with radial basis function as trail function instead of GMLS interpolation functions and test functions as simple weight function. Radial basis interpolation function yields computationally simple method involving fewer matrix inversions and multiplications. Effectiveness of proposed MLPG method was evaluated by the number of patch test and mixed boundary value problems [34]. Li et al. extended MLPG approach to three-dimensional elastostatic problems by combining two methods of MLPG family, MLPG2 and MLPG5, in order to achieve high computational efficiency. MLPG5 was applied to domain from inside to eliminate domain integration and MLPG2 was applied at nodes on boundaries and interfaces of material discontinuities so that boundary conditions and material discontinuities are satisfied. Results obtained by application of proposed technique have shown good agreement with analytical solution [35]. Han and Atluri developed MLPG method for solving three-dimensional elastodynamic problems which was derived from LSWF of the equilibrium equations by general MLPG concept and MLS shape functions. The present numerical technique imposes a correction to the accelerations to enforce the kinematic boundary conditions in MLS approximation with explicit time-integration algorithm [36]. While Long et al. proposed a new MLPG approach based on MLPG5 and coupled radial basis function (RBF) with polynomial basis function as trial function for elastodynamic problems. The shape function constructed possesses Kronecker delta property, hence no additional treatment to impose EBCs are required.

Newmark family of methods is adopted in time integration scheme and the technique does not involve any domain or singular integration [37]. As a new concept, MLPG approach with polygonal sub domains constructed from several triangular patches rather than typically used circular subdomains was presented by Pudjisuryadi. Variable domain of influence (VDOI) and effective stress gradient indicator for assessing local errors were focused in the study. VDOI helps in alleviating the problem raised by adaptive meshfree approach where problem domain is refined with new nodes placed in area where local error exceeds a level but due to constant size of influence domain, node density in that area goes too high and finally it leads to higher computational cost and ineffective adaptive technique [38]. In a recent development, a novel MLPG method with new test function, Guassian test function, as a schema to solve problems in elastostatic and fracture mechanics was developed by Abdollahifar et al. Four different variants of MLPG method, MLPG1, MLPG2, MLPG5, and MLPG6, can be approached using new test function and sufficiently accurate results were obtained [39].

Plates and Shell Problems

Due to the flexibility in constructing approximation functions with desired smoothness and accuracy in meshfree methods, they have been successfully applied to Kirchhoff type of plates and shell problems. Applications to plates were first investigated by Hein [40] but due to use of point collocation to enforce EBCs, too small supports and unsuitable weight functions desired results were not observed. Lu et al. [41] treated Mindlin-Reissner plates with linear and quadratic basis EFG method, but results were poor due to shear locking. The EFG method had been applied to thin (Kirchhoff) plates by Krysl and Belytschko, in 1995. Background quadrilateral elements were used for the purpose of Gaussian numerical integration. An attempt to optimize the accuracy of the method by the choice of the weight function support size was undertaken [42]. Same authors applied EFG method to thin shells wherein background elements were used for surface shape approximation and numerical integration. MLS was used in approximating, surface while EBCs were imposed by Lagrange multiplier. To achieve consistency, quadratic and quartic polynomial basis was used along with quartic spline weight function [43]. Liu and

Chen applied EFG method with MLS approximation to static and free vibration analysis of thin plates of complicated shapes, that is, rectangular plate, elliptical plate, and complicated shapes with different boundary conditions. In proposed approach, for static analysis EBCs were imposed by penalty method while in free vibration analysis they were imposed by orthogonal transformation technique [44]. Free vibration analysis of composite laminates of complicated shapes through EFG method was carried out by the same authors. EBCs were imposed by Lagrange multiplier and orthogonal transformation technique. Numerical examples of square plate, elliptical plate, and other complicated shapes were addressed by proposed method [45]. Dai et al. also presented a meshfree method for analyzing thin and thick laminated composite plates for static deflection and natural frequencies using higher order shear deformation theory. MLS approximants were applied to construct the shape functions and variational principle was used to derive the discrete system equations based on the third order shear deformation theory (TSDT) of Reddy. EBCs were enforced by a penalty technique for both the static deflection and natural frequency analysis [46]. The same authors presented EFG method for thermomechanical analysis of FGM plates containing distributed piezoelectric sensors and actuators with structured and unstructured (irregular) node arrangement. The weak form was formulated based on FSDT and shape functions were constructed by MLS functions. EBCs were imposed by penalty method in proposed Galerkin method. Unstructured nodes were also giving the desired order of accuracy in results [47]. Peng et al. proposed EFG method for static analysis of concentrically and eccentrically stiffened plates based on FSDT. The influences of support size and order of the complete basis function on the numerical accuracy were also investigated and it was observed that larger support size and higher order of basis function will furnish better convergence results [48]. Corrugated plates are widely used in industries due to their improved strength to weight ratio. They can be modeled and analyzed either by considering them as shells or orthotropic plates. For elastic buckling analysis of stiffened and unstiffened corrugated plates with FSDT, meshfree Galerkin method was proposed by the same authors. Corrugated plates were treated as orthotropic plates and stiffeners were taken as beams. Stiffness matrix for structure was obtained by superimposing the strain energy of the orthotropic plate and the beams and imposing the displacement compatibility conditions between the plate and the beams [49]. For analyzing elastic bending of stiffened and unstiffened corrugated plates, meshfree Galerkin method was proposed by the same authors where MLS shape functions with full transformation method was employed for enforcing EBCs [50]. The proposed method was extended for solving nonlinear problems of stiffened and unstiffened corrugated plates based on FSDT and von Karman large deformation theory. For validation of proposed approach, different corrugated plates were analyzed and results were compared with results obtained with shell elements in ANSYS [51]. Belinha and Dinis also proposed EFG method for nonlinear analysis of plates and laminates with FSDT and MLS approximants as shape functions. Laminate bending problems were solved and results were compared with FEM solution [52]. While two members of MLPG family, MLPG1 and MLPG5, were used for three-dimensional static analysis of thick functionally graded plates by Vaghefi et al. wherein MLPG1 uses fourth order spline function as test function and MLPG5 uses Heaviside step function as test function. Young's modulus is considered to be graded through the thickness of plates by an exponential function while Poission's ration is taken as constant while LSWF was derived. 3D MLS approximation was used for field variables and brick shaped domains were considered as local subdomains and support domains [53]. Mojdehi et al. also proposed MLPG for three-dimensional (3D) static and dynamic analysis of thick functionally graded plates using three-dimensional MLS shape function and Heaviside step function as test function [54]. In order to estimate maximum sustainable load by any structure, limit analysis proved to be useful technique where fundamental theorems of plastic analysis are used. For such an application h-Adaptive EFG method with MLS approximation was proposed by Le et al. for limit analysis of plates. Accuracy of limit analysis is often affected and decided by local singularities arising from localized plastic deformations and to capture it accurately automatic h-refinement is performed. Taylor expansion technique is used for error estimation in computed displacement field to identify the area needing refinement [55]. In another steel structural application, use of beams with irregular web holes of different shape and arrangement is widespread. But the behavior of such beams is complicated due to irregularity of openings and it needs more reliable technique for their local buckling response. Abidin and Izzuddin proposed EFG method for local buckling analysis of steel beams with irregular web openings. The proposed EFG approach was based on general formulation of plate buckling, where singularity in tangent stiffness matrix was made up of material stiffness matrix and geometric stiffness matrix. MLS approximation was used for shape function construction and the proposed approach was applied to three-dimensional beam buckling problems [56]. In a recent development, Jaberzadeh et al. developed EFG method for buckling analysis of inelastic skew plates with or without line supports. The governing differential equation for a plate in plastic range of response is numerically solved with Galerkin method and Stowell theory for the plastic buckling of flat skew plates with variable thickness is used. MLS approximants are used for shape function construction and EBCs are imposed by Lagrange multiplier, nd orthogonal transformation techniques in proposed method [57]. Spatial thin shell structures are used extensively in many engineering structures including aircrafts, pressure vessels, and automobiles due to its outstanding efficiency in material utilization. Liu et al. applied EFG to thin shell structures for static deformation and free vibration analysis. MLS was used for construction of shape functions and surface approximation of general spatial shell geometry and discrete system equations were obtained by incorporating these interpolations into the Galerkin weak form. EBCs were imposed by penalty approach, Lagrange multiplier, and orthogonal transformation techniques [58]. A meshfree method for static analysis of FGM cylinders was presented by Foroutan et al., wherein mechanical properties were assumed to vary in radial direction. EBCs were imposed by transformation method and MLS approximants were used for approximating unknown filed variable. The method was applied to finite and infinite length cylinders and results obtained were in good agreement with FEM [59].

Large Deformation and Contact Problems

Numerical simulation of contact between two different objects like sheet metal forming, vehicle crash worthiness, impact, penetration, and so forth, is a challenging task. While dealing with contact problems, mesh density must be maintained at a sufficiently high level around contact region to obtain reasonably accurate results. Li et al. proposed a contact detection algorithm based on moment matrix of meshfree approximation. The mathematical principle of contact detection algorithm is that the determinant of moment matrix can automatically determine Lagrangian movement of continuum and on the basis of that one can accurately detect contact or penetration without solving any complex equations. It was implemented to simulate Taylor bar impact problem which is a deformable solid bar impacting rigid target problem [60]. Large strain problems, like hyperelastic materials undergoing large deformations, cannot be handled with ease in FEM due to excessive mesh distortion, but meshfree methods proved to be a good alternative in those cases. Tiago and Pimenta implemented EFG with MLS approximant to nonlinear analysis of plates undergoing arbitrary large deformations which is based on a unified nonlinear theory of plates allowing arbitrarily large rotations and displacements. Presented approach was hybrid in nature where solution was obtained by the independent approximation of the generalized internal displacement fields and generalized boundary tractions [61]. Hu et al. developed MLPG approach for large deformation contact analysis of elastomers like rubber block and compression of rubber ring. Proposed MLPG approach was based on a local weak form with RBF coupled with polynomial basis function. In the present technique, two different sets of equations were used for nodes on the contact surface and nodes away from contact surface [62]. Li and Lee developed an adaptive meshless method, with sliding line algorithm and penalty method to handle contact constraints, for solving contact problems involving large deformation in which additional nodes are added automatically into large error regions. For automatic node insertion, a modified error estimation (built on two different support sizes of a basis function) was proposed to identify regions of large computational errors [63]. A novel complex variable EFG method, improved complex variable EFG, for two-dimensional large deformation problem was developed by Li et al. Based on complex variable theory and moving least-squares (MLS) approximation, the improved complex variable moving least-squares (ICVMLS) approximation was developed. Proposed technique was based on Galerkin weak form, while the penalty method was used to impose EBCs [64]. A recent development in nonlinear solid mechanics is proposed by Ullah and Augarde by developing adaptive meshless approach based on EFG method. An existing error estimation procedure for linear elastostatic problems is extended for nonlinear problems including finite deformation and elastoplasticity. Proposed max-ent EFG method can handle material and geometrically nonlinear problems in solid mechanics including a robust means of transferring data between discretizations [65].

Sensitivity Analysis and Shape Optimization

Design sensitivity analysis and shape optimization is applied to observe and find the variation of response measure like displacement, stress due to variation of some design parameter that is, geometric parameter, and finding optimal layout of a structure within a specified region. Bobaru and Mukherjee demonstrated application of EFG to shape design sensitivity analysis and shape optimization for 2D elasticity problems wherein EFG was used first time with continuous formulation using material derivative approach. Penalty approach was used for imposing EBCs and a numerical example of shape optimization of fillet was used to demonstrate robustness and ability of EFG method. The presented approach can be extended to 3D and nonlinear problems. Another application of EFG to shape optimization in linear thermoelasticity problem was also demonstrated by the same authors [66, 67]. Zhang et al. proposed meshless computational strategies for shape optimal design through the composition of behavioral fields quite similar to Boolean operations in constructive solid geometry (CSG). A meshless approximation using nonuniform rational B-spline basis functions was used to discretize the behavioral fields defined over the geometrical primitives while remeshing was performed for only those primitives that were modified. Due to a tighter integration between design and analysis, it is termed constructive solid analysis (CSA) [68]. Juan et al. introduced a technique to combine EFG with evolutionary structural optimization (ESO) for topology optimization of the continuum structures where objective function of the model is to minimize weight by gradually removing the inefficient material from the design domain. Feasibility and efficiency of the proposed technique was illustrated with several 2D examples like cantilever beams and simply supported beams [69].

FRACTURE MECHANICS

Static and Dynamic Fracture

Belytschko et al. developed EFG method for linear elastic fracture problems in 1994 where "Visibility criterion" was proposed firstly to model geometric discontinuity (crack) in which domain of influence for nodes near the crack are truncated whenever they intersect the crack surface and hence a node on one side of crack will not affect the point on other side of crack. But "visibility criterion" had difficulty in treating nodes near crack tip. Other improved techniques for handling discontinuity were also suggested: diffraction method, transparency method, and "see through" method or continuous line criterion. Two methods for enriching EFG approximations for linear elastic fracture problems were also proposed by Belytschko: extrinsic and intrinsic enrichment [70, 71]. Belytschko and Tabbara proposed EFG approach for dynamic fracture problem in 1996 involving numerical examples of crack propagation at constant velocity and constant value of dynamic fracture toughness [72], while for dynamic propagation of arbitrary three-dimensional cracks, EFG approach was developed by Krysl and Belytschko. MLS shape function and truncated Gaussian weight functions were employed in proposed EFG approach [73]. Belytschko et al. studied mixed mode dynamic crack growth in concrete using EFG methods wherein fracture process zone (FPZ) technique was used in formulation as linear elastic fracture mechanics (LEFM) is not applicable to concrete and other cement-based materials. Numerical examples of mode I and mixed mode cracks in concrete were discussed [74]. Rao and Rahman proposed an efficient meshfree method, based on EFG, for linear elastic crack problems with single and mixed mode loading conditions. Proposed technique involves a new weight function and exact implementation of EBCs. Estimated SIFs and neat tip stress fields were in good agreement with FEM results and experimental values [75]. Tiago and Leitão also applied EFG to damage analysis of reinforced concrete beams which demonstrated handling material inhomogeneities, discontinuity in geometry, and concentrated loads [76]. Lee and Yoon proposed an enhanced EFG method with enhancement functions to improve solution efficiency for linear elastic fracture problems where singularity and discontinuity of crack were modeled with enhancement function and discontinuity functions, respectively. EBCs were imposed by penalty method and coupling with FEM [77]. Rabczuk and Belytschko proposed a new EFG approach, called EFG-Particle (EFG-P), for modeling discrete cracks wherein crack growth was represented by activation of crack surfaces at individual particles and hence crack's topology representations is not needed. The crack was modeled by local enrichment of trial and test functions with sign function and it can handle crack branching and fragmentation also [78]. One of the most efficient MLPG variant, MLPG5, was used for analysis of elastodynamic deformations near crack tip by Kaiyuan et al. Newmark family of methods was applied into the time integration scheme. A numerical example of a rectangular plate with a parallel central crack loaded in tension was approached by proposed technique [79]. EFG method was extended to solve three-dimensional elastic fracture mechanics problems, mode I and mode II cracks, by Brighenti. Out of different ways for modeling geometric discontinuity by EFG method, "visibility criterion" was used to detect them in present approach, while Gauss type weight function along with penalty approach employed to enforce the boundary conditions. Presented approach was validated by solving thick plate with an edge crack under tension and finite thin plate with central slant crack under tension [80]. Improved element free galerkin method (IEFG) with an improved moving least square approximation (IMLS) was developed for analyzing two-dimensional fracture mechanics problems by Zhang et al. Major advantage with IMLS is its greater computational efficiency than MLS and it does not lead to ill-conditioned system of equations as MLS does sometimes [81]. The effective and accurate calculation of stress intensity factors (SIF) is one of the basic problems in LEFM. Two principal approaches are known for a SIF calculation: local, based on the use of displacements or tractions near to the crack tip; and global or energy methods, based on the calculation of the energy release rate in terms of crack growing. Parvanova presented a procedure for calculation of SIFs based on standard appearance of force-displacement curve using EFG method. The method was used to develop a new idea based on standard appearance of the force-displacement curve in LEFM related to the accurate derivation of the SIF for pure opening mode I type fracture. A MATLAB code for two-dimensional elasticity problems had been worked out, along with intrinsic basis enrichment for precise modeling of the singular stress field around the crack tip [82]. Zhang and Chen proposed a simplified meshfree method with Kronecker delta property for incorporation of displacement boundary conditions for dynamic crack growth problem wherein the crack was presented by a set of rotated crack segments that pass through the entire domain of influence of the meshfree nodes. Rankine criterion was used to initiate the crack and discontinuous displacement field was obtained by an extrinsic enrichment based on a local partition of unity concept [83]. A meshfree analysis of dynamic fracture in thin walled structures was proposed by Gato, where fracture of the shell is modeled by breaking links between particles once a certain fracture criterion is met. For validating the proposed approach, numerical examples of quasistatic tearing of a square plate, an impact problem and detonation driven fracture of cylindrical shells were considered. In present work, it was the first time that the 3D continuum approach based on Lagrangian kernels was applied to fracture of thin shells and implementation was done in C++ [84, 85]. In a recent development, a new enrichment criterion for modeling kinked cracks using EFG method is proposed by Pant et al. In order to capture crack tip stress singularity, some additional terms are incorporated in the linear basis function. The proposed criterion is applied for simulating the quasistatic crack growth in two-dimensional domain subjected to mixed mode loading [86].

Composite Solids

Functionally graded materials (FGM) possesses continuously varying microstructure and material properties in a predetermined way and they are used in structures subjected to nonuniform service conditions. Rao and Rahman proposed EFG approach for calculating stress intensity factors (SIF) for stationary crack in twodimensional functionally graded materials of arbitrary geometry. In proposed method, the interaction integral method was extended for FGM and material properties were taken as smooth functions of spatial coordinates and two newly developed interaction integrals were introduced for analysis of basic modes and mixed mode fracture problems [87]. While two-dimensional stress analysis problems of anisotropic and linear elastic/viscoelastic solids with continuously varying material properties were addressed by Sladek et al., using MLPG with unit step function as test function in local weak form which leads to local boundary integral equations (LBIEs). MLS was adopted for approximating physical quantities in LBIEs and for time-dependent problems, Laplace transformation was utilized [88]. Delamination and matrix cracking are routine damage mechanisms observed during analysis of laminated structures. Guiamatsia et al. proposed EFG for the first time to simulate delamination (interlaminar) and intralaminar matrix microcracking in composite laminates. Modeling was done at mesolevel, where each ply represented individually and background integration cells were arranged per layer in a way that they are not traversed by material interfaces. Virtual crack closure technique (VCCT) was used for crack advancement in the present technique [89]. Orthotropic composites possess high specific strength and stiffness characteristics because of their constituents and extensively applied in various engineering applications. Ghorashi et al. presented a new approach for modeling discrete cracks in two-dimensional orthotropic media by EFG method. In proposed approach, recently developed orthotropic enrichment functions were used which were used earlier in the extended finite element method along with a subtriangle technique for enhancing the Gauss quadrature accuracy near the crack [90].

SOME TYPICAL APPLICATIONS AND ENHANCEMENT OF COMPUTATIONAL EFFICIENCY

Typical Applications

Over a period of time in service, mechanical systems and structures accumulate cracks due to fatigue. Duflot and Nguyen-Dang proposed an enriched meshless method to analyze fatigue crack growth under cyclic loading. The crack propagation was modeled by successive linear extensions determined by SIFs obtained after each linear elastic analysis. A fixed set of three nodes with special weight function were added at each crack tip to accurately catch the stress singularities in proposed approach [91]. Use of modal characteristics, that is, vibration data, like natural frequency and mode shapes, of structures for detecting and predicting cracks has become a good alternative approach because cracked structure's modal data will be different.

In order to use vibration data for detecting cracks in load carrying systems or structures, two different theoretical modeling techniques are used: lumped flexibility models and continuous models. Andreaus et al. proposed MLPG approach with MLS shape function for analyzing vibration of beams with multiple cracks. Lumped flexibility model was adopted and each fatigue crack was modeled as rotational spring in proposed approach [92]. In impact problems, transition in failure mode can be observed. But most numerical simulation techniques focus either on brittle failure or ductile failure. Kalthoff and Winkler conducted experiments on prenotched specimen of steel plates, subjected to impact loading with different impact velocities and observed transition in failure modes. Wang and Liu proposed EFG method for simulating failure transition from brittle to ductile under finite deformation. Johnson-Cook damage model was incorporated in Galerkin formulation and EBCs were enforced by collocation method. Node splitting algorithm was used in modeling crack which simplifies the implementation. Proposed method captured the complicated failure transition phenomenon accurately [93]. Under combined mechanical and thermal loadings, the presence of cracks induces a strong variation in fields, which can affect the crack growth direction. While designing structures for turbines, combustion chambers, and nuclear pressure vessels, thermoelastic fracture mechanics aspects need to be considered. Pant et al. proposed intrinsic enriched EFG to solve thermoelastic fracture mechanics problem in homogeneous and inhomogeneous materials (bi-material). The thermo-elastic fracture problem was solved by decoupling it into two separate problems and both the problems were enriched intrinsically to represent discontinuous temperature, heat flux, displacements, and traction across crack surfaces. For modeling bi-material interface, jump function technique had been employed in proposed method [94]. Extended finite element method (XFEM) was developed to ease difficulties in solving problems with geometric discontinuity like cracks by adding discontinuous basis function to standard polynomial basis functions for nodes that belonged to elements intersected by crack. In a recent development, advantages of a meshfree method—EFG are combined with XFEM and extended element free Galerkin method (XEFG) is proposed to model crack propagation under thermomechanical loading by Bouhala et al.. In proposed method, direction of the crack growth is determined by initially calculating SIFs using the interaction energy integral, and then the crack is assumed to propagate in the direction of the maximum principal stress. Shape functions are constructed using MLS approximation and cracks; interfaces and crack tips are modeled with extrinsic local enrichment [95].

Enhancing Computational Efficiency and Error Control

Meshfree shape functions are not interpolation functions and do not possess Kronecker delta properties. Hence imposition of EBCs consumes much computation time. Several techniques were presented for imposing EBCs, like Lagrange multiplier, penalty method, orthogonal transformation techniques, coupling with FEM, Nitsche's method, singular weighing functions, boundary collocation, and D'Alembert's Principle. An overview of existing techniques for enforcing EBCs was presented by Fernández-Méndez and Huerta. With focus on meshfree method coupled with finite elements and methods based on modification of Galerkin weak form [17]. Chen and Wang proposed two new boundary condition treatment techniques, the mixed transformation method and the boundary singular kernel method, to enhance the computational efficiency of meshfree methods for contact problems in RKPM framework. The mixed transformation method is a modification of a full transformation method developed previously for meshfree solution of boundary value problems, while the boundary singular kernel method introduces singularities into the kernel functions associated with the restrained nodes [18]. Another set of new boundary treatment techniques were developed by Ren and Liew, namely, node interpolation method (NIM) and direct imposition method (DIM). In NIM, the shape functions associated with EBCs were constructed using node interpolation and then combined with meshfree shape functions, while DIM rearranges the discretized system equations, and directly provides the known values of the essential boundary conditions in the nodal variable vector [19]. Smoothing of the approximating functions at concave boundaries and accelerated calculations of the approximating function in EFG method was proposed by Belytschko et al. In proposed method, shape functions of EFG method were modified and made continuous in domain with concave corners, by simply redefining a parameter governing decay of weight function [96]. In meshfree methods, for numerical integration of Galerkin weak form, Gauss integration method is most commonly used. Dolbow and Belytschko demonstrated and investigated integration aspects in meshfree methods. Authors emphasized on source of integration errors and suggested techniques to minimize them [20]. But a number of disadvantages have been reported in employing Gauss integration, like complexity, requirement of higher order quadrature rules, specialized integration zone patterns, and so forth. Beissel and Belytschko used direct nodal integration to avoid background cells, but it led to oscillations in solution due to underintegration of weak form and vanishing shape functions at nodes. To overcome it, a stabilized nodal integration technique, for EFG, was proposed by adding a residual of the equilibrium equation to the potential energy function. The proposed technique does not need background cell structure and results in completely meshless method [21]. A stabilized conforming nodal integration (SCNI) method for elastoplastic contact analysis of metal forming processes was proposed by Yoon et al. In this approach, strain smoothing stabilization was introduced to eliminate spatial instability in collocation meshfree methods and convergence was obtained by introducing an integration constraint (IC) as a necessary condition for a linear exactness in the mesh-free Galerkin approximation. Implementation of proposed technique in linear problems demonstrated a significant reduction in computational cost with no loss of accuracy and convergent rate compared to the solution obtained by the use of Gauss integration [22, 97]. In an another application of meshfree formulation with SCNI, Wang and Chen proposed locking free meshfree curved beam formulation based on SCNI with Kirchhoff mode reproducing conditions (KRMC). Proposed meshfree approximation was constructed to represent pure bending mode without producing parasitic shear and membrane deformations. Numerical examples of pure bending of clamped-free curved beam, a nearly straight beam with tip load and a pinched ring demonstrated the technique [98]. Khosravifard and Hematiyan presented a technique for evaluation of regular domain integrals without domain discretization wherein a domain integral is transformed into a boundary integral and a 1D integral and then utilized for domain integrals in meshfree methods based on weak form like EFG method. Presented technique results in truly meshless approach with better accuracy and efficiency. It is known as Cartesian transformation method (CTM) which was used earlier for domain integration in boundary element method by Hematiyan computations, and so forth, Chung [99].

Though meshfree methods look quite attractive for solving a special class of problems, there are issues like error estimate and control, integral evaluation, and accuracy. Errors in numerical modeling arise from number of sources like discretization, quality of mathematical model, rounding off operations in computations, and so forth. Chung and Belytschko, in 1998, proposed estimation of local and global error in EFG method where in error estimation was based on difference between the values of projected stress and stress given by EFG method. Effectiveness of proposed error estimator was validated by various one-dimensional and two-dimensional problems [100]. Zhuang et al. investigated discretization error in EFG method. Discretization errors are arising due to not satisfying governing equation and boundary conditions. Conventional procedures for error analysis used in FEM cannot be applied straightly in meshfree approaches. In FEM, it is feasible to uncouple h and p adaptivity but in EFG method it is not possible because changing the density of nodes both changes the error e_h and also changes the space of the shape functions and hence the error e_p. Hence it is difficult to achieve error control and adaptivity in meshfree methods. In proposed approach the discretization error was split into contributions arising from an inadequate number of degrees of freedom e_h and from an inadequate basis e_p [101]. Kim and Atluri proposed a technique for controlling error and improving solution accuracy in MLPG method by adding and arbitrary placing secondary nodes where better resolution is needed in the domain. But the subdomains for the shape functions in the MLS approximation were constructed only from the primary nodes, and the secondary nodes use the same sub-domains. The proposed technique can become very useful in an adaptive approach, because the secondary nodes can be easily added and/or moved without an additional mesh [102].

CONCLUSION AND DISCUSSIONS

Objective of the present work is to provide exposure in terms of versatility of meshfree methods in handling different types of engineering problems without detailed mathematical description. Some of the worth notable recent developments include development of extended element free Galerkin (XEFG) method for thermo-mechanical crack propagation, CVEFG using CVMLS for elastoplasticity problems, ICVEFG using ICVMLS for large deformation problems, simulation of failure transition in impact problems by EFG, local buckling analysis of steel plates with irregular openings by EFG, and adaptive EFG and MLPG approaches.

Though MMs have found application in almost all areas of structure and fracture mechanics, still there are challenges in developing computationally efficient algorithms with accurate nodal integration techniques with scalable implementation of EBCs. Improved versions of original MMs are also proposed by many researchers like XEFG, IEFG, CVEFG, ICVEFG, and so forth, while some of the other techniques for improving computational efficiency are reviewed in Section 5. MMs have exemplary approximation, but the computational cost is the issue. To alleviate the problem, coupled numerical methods like FEM-MM have been developed to exploit potential benefits of each method.

REFERENCES

1. G. R. Liu and M. B. Liu, Smoothed Particle Hydrodynamics: A Meshfree Particle Method, World Scientific Publishing, Singapore, 2003.
2. L. B. Lucy, "A numerical approach to the testing of the fission hypothesis," Astronomical Journal, vol. 82, pp. 1013–1024, 1977.
3. R. A. Gingold and J. J. Monaghan, "Smoothed particle hydrodynamics: theory and allocation to non-spherical stars," Monthly Notices of the Royal Astronomical Society, vol. 181, pp. 375–389, 1977.
4. J. J. Monaghan, "An introduction to SPH," Computer Physics Communications, vol. 48, no. 1, pp. 89–96, 1988.
5. T. Belytschko, Y. Y. Lu, and L. Gu, "Element-free Galerkin methods," International Journal for Numerical Methods in Engineering, vol. 37, no. 2, pp. 229–256, 1994.
6. K. M. Liew, Y. Cheng, and S. Kitipornchai, "Boundary element-free method (BEFM) and its application to two-dimensional elasticity problems," International Journal for Numerical Methods in Engineering, vol. 65, no. 8, pp. 1310–1332, 2006.

7. W. K. Liu, S. Jun, and Y. F. Zhang, "Reproducing kernel particle methods," International Journal for Numerical Methods in Fluids, vol. 20, no. 8-9, pp. 1081–1106, 1995.
8. G. R. Liu, Meshfree Methods: Moving beyond the Finite Element Method, CRC Press, Taylor & Francic Group, 2nd edition, 2010.
9. T. Belytschko, Y. Krongauz, D. Organ, M. Fleming, and P. Krysl, "Meshless methods: an overview and recent developments," Computer Methods in Applied Mechanics and Engineering, vol. 139, no. 1–4, pp. 3–47, 1996.
10. S. Li and W. K. Liu, "Meshfree and particle methods and their applications," Applied Mechanics Reviews, vol. 55, no. 1, pp. 1–34, 2002.
11. V. P. Nguyen, T. Rabczuk, S. Bordas, and M. Duflot, "Meshless methods: a review and computer implementation aspects," Mathematics and Computers in Simulation, vol. 79, no. 3, pp. 763–813, 2008.
12. K. M. Liew, X. Zhao, and A. J. M. Ferreira, "A review of meshless methods for laminated and functionally graded plates and shells," Composite Structures, vol. 93, no. 8, pp. 2031–2041, 2011.
13. S. N. Atluri and T. L. Zhu, "A new meshless local Petrov-Galerkin (MLPG) approach in computational mechanics," Computational Mechanics, vol. 22, no. 2, pp. 117–127, 1998.
14. Y. Y. Lu, T. Belytschko, and L. Gu, "A new implementation of the element free Galerkin method,"Computer Methods in Applied Mechanics and Engineering, vol. 113, no. 3-4, pp. 397–414, 1994.
15. Y. Krongauz and T. Belytschko, "Enforcement of essential boundary conditions in meshless approximations using finite elements," Computer Methods in Applied Mechanics and Engineering, vol. 131, no. 1-2, pp. 133–145, 1996.
16. F. C. Günther and W. K. Liu, "Implementation of boundary conditions for meshless methods,"Computer Methods Appllied Mechanics and Engineering, vol. 163, no. 1–4, pp. 205–230, 1998.
17. S. Fernández-Méndez and A. Huerta, "Imposing essential boundary conditions in mesh-free methods," Computer Methods Applied Mechanics and Engineering, vol. 193, no. 12–14, pp. 1257–1273, 2004.

18. J. S. Chen and H. P. Wang, "New boundary condition treatments in meshfree computation of contact problems," Computer Methods in Applied Mechanics and Engineering, vol. 187, no. 3-4, pp. 441–468, 2000.
19. J. Ren and K. M. Liew, "Mesh-free method revisited: two new approaches for the treatment of essential boundary conditions," International Journal of Computational Engineering Science, vol. 3, no. 2, pp. 219–233, 2002.
20. J. Dolbow and T. Belytschko, "Numerical integration of the Galerkin weak form in meshfree methods," Computational Mechanics, vol. 23, no. 3, pp. 219–230, 1999.
21. S. Beissel and T. Belytschko, "Nodal integration of the element-free Galerkin method," Computer Methods in Applied Mechanics and Engineering, vol. 139, no. 1–4, pp. 49–74, 1996.
22. J. S. Chen, C. T. Wu, S. Yoon, and Y. You, "A stabilized conforming nodal integration for Galerkin mesh-free methods," International Journal for Numerical Methods Engineering, vol. 50, no. 2, pp. 435–466, 2001.
23. P. Krysl and T. Belytschko, "ESFLIB: a library to compute the element free Galerkin shape functions," Computer Methods in Applied Mechanics and Engineering, vol. 190, no. 15–17, pp. 2181–2205, 2001.
24. J. Dolbow and T. Belytschko, "An introduction to programming the meshless element free Galerkin method," Archives Computational Methods in Engineering, vol. 5, no. 3, pp. 207–241, 1998.
25. W. Barry and S. Saigal, "A three-dimensional element-free Galerkin elastic and elastoplastic formulation," International Journal for Numerical Methods in Engineering, vol. 46, no. 5, pp. 671–693, 1999.
26. C. Tiago and V. M. A. Leitão, "Analysis of free vibration problems with the element-free Galerkin method," in Proceedings of the 9th International Conference on Numerical Methods in Continuum Mechanics, Žilina, Slovakia, September 2003.
27. Z. Zhang, P. Zhao, and K. M. Liew, "Improved element-free Galerkin method for two-dimensional potential problems," Engineering Analysis with Boundary Elements, vol. 33, no. 4, pp. 547–554, 2009.

28. Z. Zhang, P. Zhao, and K. M. Liew, "Analyzing three-dimensional potential problems with the improved element-free Galerkin method," Computational Mechanics, vol. 44, no. 2, pp. 273–284, 2009.
29. M. Peng, D. Li, and Y. Cheng, "The complex variable element-free Galerkin (CVEFG) method for elasto-plasticity problems," Engineering Structures, vol. 33, no. 1, pp. 127–135, 2011.
30. S. N. Atluri and T. L. Zhu, "Meshless local Petrov-Galerkin (MLPG) approach for solving problems in elasto-statics," Computational Mechanics, vol. 25, no. 2, pp. 169–179, 2000.
31. S. N. Atluri and S. Shen, "The meshless local Petrov-Galerkin (MLPG) method: a simple & less-costly alternative to the finite element and boundary element methods," Computer Modeling in Engineering and Sciences, vol. 3, no. 1, pp. 11–51, 2002.
32. S. Long and S. N. Atluri, "A meshless local Petrov-Galerkin method for solving the bending problem of a thin plate," Computer Modeling in Engineering and Sciences, vol. 3, no. 1, pp. 53–63, 2001.
33. I. S. Raju and D. R. Phillips, A Meshless Local Petrov-Galerkin Method for Euler-Bernoulli Beam Problems, NASA Langley Research Center, Hampton, Va, USA, 2001.
34. I. S. Raju, D. R. Phillips, and T. Krishnamurthy, Meshless Local-Petrov Galerkin Euler-Bernoulli Beam Problems: A Radial Basis Function Approach, NASA Langley Research Center, Hampton, Va, USA, 2001.
35. Q. Li, S. Shen, Z. D. Han, and S. N. Atluri, "Application of meshless local Petrov-Galerkin (MLPG) to problems with singularities, and material discontinuities, in 3-D elasticity," Computer Modeling in Engineering and Sciences, vol. 4, no. 5, pp. 571–585, 2003.
36. Z. D. Han and S. N. Atluri, "A Meshless Local Petrov-Galerkin (MLPG) approach for 3-dimensional elasto-dynamics," Computers, Materials and Continua, vol. 1, no. 2, pp. 129–140, 2004.
37. S. Y. Long, K. Y. Liu, and D. A. Hu, "A new meshless method based on MLPG for elastic dynamic problems," Engineering Analysis with Boundary Elements, vol. 30, no. 1, pp. 43–48, 2006.

38. P. Pudjisuryadi, "Adaptive meshless local Petrov-Galerkin method with variable domain of influence in 2D elastostatic problems," Civil Engineering Dimension, vol. 10, no. 2, pp. 99–108, 2008.
39. A. Abdollahifar, M. R. Nami, and A. R. Shafiei, "A new MLPG method for elastostatic problems,"Engineering Analysis with Boundary Elements, vol. 36, no. 3, pp. 451–457, 2012.
40. P. Hein, "Diffuse element method applied to Kirchhoff plates," Technical Report, Department of Civil Engineering, Northwestern University, Evanston, Ill, USA, 1993.
41. Y. Y. Lu, L. Gu, and T. Belytschko, "Diffuse element method applied to Kirchhoff plates," Internal report, 1996.
42. P. Krysl and T. Belytschko, "Analysis of thin plates by the element-free Galerkin method,"Computational Mechanics, vol. 17, no. 1-2, pp. 26–35, 1995.
43. P. Krysl and T. Belytschko, "Analysis of thin shells by the element-free Galerkin method,"International Journal of Solids and Structures, vol. 33, no. 20–22, pp. 3057–3080, 1996.
44. G. R. Liu and X. L. Chen, "A mesh-free method for static and free vibration analyses of thin plates of complicated shape," Journal of Sound and Vibration, vol. 241, no. 5, pp. 839–855, 2001.
45. X. L. Chen, G. R. Liu, and S. P. Lim, "An element free Galerkin method for the free vibration analysis of composite laminates of complicated shape," Composite Structures, vol. 59, no. 2, pp. 279–289, 2003.
46. K. Y. Dai, G. R. Liu, K. M. Lim, and X. L. Chen, "A mesh-free method for static and free vibration analysis of shear deformable laminated composite plates," Journal of Sound and Vibration, vol. 269, no. 3–5, pp. 633–652, 2004.
47. K. Y. Dai, G. R. Liu, X. Han, and K. M. Lim, "Thermomechanical analysis of functionally graded material (FGM) plates using element-free Galerkin method," Computers and Structures, vol. 83, no. 17-18, pp. 1487–1502, 2005.
48. L. X. Peng, S. Kitipornchai, and K. M. Liew, "Analysis of rectangular stiffened plates under uniform lateral load based on FSDT and element-free Galerkin method," International Journal of Mechanical Sciences, vol. 47, no. 2, pp. 251–276, 2005.

49. K. Liew, L. Peng, and S. Kitipornchai, "Buckling analysis of corrugated plates using a mesh-free Galerkin method based on the first-order shear deformation theory," Computational Mechanics, vol. 38, no. 1, pp. 61–75, 2006.
50. L. X. Peng, K. M. Liew, and S. Kitipornchai, "Analysis of stiffened corrugated plates based on the FSDT via the mesh-free method," International Journal of Mechanical Sciences, vol. 49, no. 3, pp. 364–378, 2007.
51. K. M. Liew, L. X. Peng, and S. Kitipornchai, "Nonlinear analysis of corrugated plates using a FSDT and a meshfree method," Computer Methods in Applied Mechanics and Engineering, vol. 196, no. 21–24, pp. 2358–2376, 2007.
52. J. Belinha and L. M. J. S. Dinis, "Nonlinear analysis of plates and laminates using the element free Galerkin method," Composite Structures, vol. 78, no. 3, pp. 337–350, 2007.
53. R. Vaghefi, G. H. Baradaran, and H. Koohkan, "Three-dimensional static analysis of thick functionally graded plates by using meshless local Petrov-Galerkin (MLPG) method," Engineering Analysis with Boundary Elements, vol. 34, no. 6, pp. 564–573, 2010.
54. A. R. Mojdehi, A. Darvizeh, A. Basti, and H. Rajabi, "Three dimensional static and dynamic analysis of thick functionally graded plates by the meshless local Petrov-Galerkin (MLPG) method,"Engineering Analysis with Boundary Elements, vol. 35, no. 11, pp. 1168–1180, 2011.
55. C. V. Le, H. Askes, and M. Gilbert, "Adaptive element-free Galerkin method applied to the limit analysis of plates," Computer Methods in Applied Mechanics and Engineering, vol. 199, no. 37–40, pp. 2487–2496, 2010.
56. A. R. Z. Abidin and B. A. Izzuddin, "Meshless local buckling analysis of steel beams with irregular web openings," Engineering Structures, vol. 50, pp. 197–206, 2013.
57. E. Jaberzadeh, M. Azhari, and B. Boroomand, "Inelastic buckling of skew and rhombic thin thickness-tapered plates with and without intermediate supports using the element-free Galerkin method,"Applied Mathematical Modelling, vol. 37, no. 10-11, pp. 6838–6854, 2013.

58. L. Liu, G. R. Liu, and V. B. C. Tan, "Element free method for static and free vibration analysis of spatial thin shell structures," Computer Methods in Applied Mechanics and Engineering, vol. 191, no. 51-52, pp. 5923–5942, 2002.
59. M. Foroutan, R. Moradi-Dastjerdi, and R. Sotoodeh-Bahreini, "Static analysis of FGM cylinders by a mesh-free method," Steel and Composite Structures, vol. 12, no. 1, pp. 1–11, 2012.
60. S. Li, D. Qian, W. K. Liu, and T. Belytschko, "A meshfree contact-detection algorithm," Computer Methods in Applied Mechanics and Engineering, vol. 190, no. 24-25, pp. 3271–3292, 2001.
61. C. Tiago and P. M. Pimenta, "An EFG method for the nonlinear analysis of plates undergoing arbitrarily large deformations," Engineering Analysis with Boundary Elements, vol. 32, no. 6, pp. 494–511, 2008.
62. D. A. Hu, S. Y. Long, X. Han, and G. Y. Li, "A meshless local Petrov-Galerkin method for large deformation contact analysis of elastomers," Engineering Analysis with Boundary Elements, vol. 31, no. 7, pp. 657–666, 2007.
63. Q. Li and K. M. Lee, "An adaptive meshless method for analyzing large mechanical deformation and contacts," Journal of Applied Mechanics, vol. 75, no. 4, Article ID 041014, 2008.
64. D. Li, B. Bai, Y. Cheng, and K. M. Liew, "A novel complex variable element-free Galerkin method for two-dimensional large deformation problems," Computer Methods Applied Mechanics and Engineering, vol. 233–236, pp. 1–10, 2012.
65. Z. Ullah and C. E. Augarde, "Finite deformation elasto-plastic modelling using an adaptive meshless method," Computers and Structures, vol. 118, pp. 39–52, 2013.
66. F. Bobaru and S. Mukherjee, "Shape sensivity analysis and shape optimization in planar elasticity using th element-free Galerkin method," Computer Methods in Applied Mechanics and Engineering, vol. 190, no. 32-33, pp. 4319–4337, 2001.
67. F. Bobaru and S. Mukherjee, "Meshless approach to shape optimization of linear thermoelastic solids," International Journal for Numerical Methods in Engineering, vol. 53, no. 4, pp. 765–796, 2002.

68. X. Zhang, M. Rayasam, and G. Subbarayan, "A meshless, compositional approach to shape optimal design," Computer Methods in Applied Mechanics and Engineering, vol. 196, no. 17–20, pp. 2130–2146, 2007.
69. Z. Juan, L. Shuyao, and L. Guangyao, "The topology optimization design for continuum structures based on the element free Galerkin method," Engineering Analysis with Boundary Elements, vol. 34, no. 7, pp. 666–672, 2010.
70. T. Belytschko, L. Gu, and Y. Y. Lu, "Fracture and crack growth by element free Galerkin methods," Modelling and Simulation in Materials Science and Engineering, vol. 2, no. 3, article 007, pp. 519–534, 1994.
71. T. Belytschko and M. Fleming, "Smoothing, enrichment and contact in the element-free Galerkin method," Computers and Structures, vol. 71, no. 2, pp. 173–195, 1999.
72. T. Belytschko and M. Tabbara, "Dynamic fracture using element-free Galerkin methods," International Journal for Numerical Methods in Engineering, vol. 39, no. 6, pp. 923–938, 1996.
73. P. Krysl and T. Belytschko, "The element free Galerkin method for dynamic propagation of arbitrary 3-D cracks," International Journal for Numerical Methods in Engineering, vol. 44, no. 6, pp. 767–800, 1999.
74. T. Belytschko, D. Organ, and C. Gerlach, "Element-free Galerkin methods for dynamic fracture in concrete," Computer Methods in Applied Mechanics and Engineering, vol. 187, no. 3-4, pp. 385–399, 2000.
75. B. N. Rao and S. Rahman, "Efficient meshless method for fracture analysis of cracks," Computational Mechanics, vol. 26, no. 4, pp. 398–408, 2000.
76. C. M. Tiago and V. M. A. Leitão, "Development of a EFG formulation for damage analysis of reinforced concrete beams," Computers and Structures, vol. 82, no. 17–19, pp. 1503–1511, 2004.
77. S. H. Lee and Y. C. Yoon, "Numerical prediction of crack propagation by an enhanced element-free Galerkin method," Nuclear Engineering and Design, vol. 227, no. 3, pp. 257–271, 2004.

78. T. Rabczuk and T. Belytschko, "Cracking particles: a simplified meshfree method for arbitrary evolving cracks," International Journal for Numerical Methods in Engineering, vol. 61, no. 13, pp. 2316–2343, 2004.
79. L. Kaiyuan, L. Shuyao, and L. Guangyao, "A simple and less-costly meshless local Petrov-Galerkin (MLPG) method for the dynamic fracture problem," Engineering Analysis with Boundary Elements, vol. 30, no. 1, pp. 72–76, 2006.
80. R. Brighenti, "Application of the element-free Galerkin meshless method to 3-D fracture mechanics problems," Engineering Fracture Mechanics, vol. 72, no. 18, pp. 2808–2820, 2005.
81. Z. Zhang, K. M. Liew, Y. Cheng, and Y. Y. Lee, "Analyzing 2D fracture problems with the improved element-free Galerkin method," Engineering Analysis with Boundary Elements, vol. 32, no. 3, pp. 241–250, 2008.
82. S. Parvanova, "Calculation of stress intensity factors based on force-displacement curve using element free Galerkin method," Journal of Theoretical and Applied Mechanics, vol. 42, no. 1, pp. 23–40, 2012.
83. Y. Y. Zhang and L. Chen, "Impact simulation using simplified meshless method," International Journal of Impact Engineering, vol. 36, no. 5, pp. 651–658, 2009.
84. C. Gato, "Meshfree analysis of dynamic fracture in thin-walled structures," Thin-Walled Structures, vol. 48, no. 3, pp. 215–222, 2010.
85. C. Gato and Y. Shie, "Dynamic analysis of fracture in thin-walled pipes," Journal of Pressure Vessel Technology, vol. 133, no. 6, Article ID 064501, 2011.
86. M. Pant, I. V. Singh, and B. K. Mishra, "A novel enrichment criterion for modeling kinked cracks using element free Galerkin method," International Journal of Mechanical Sciences, vol. 68, pp. 140–149, 2013.
87. B. N. Rao and S. Rahman, "Mesh-free analysis of cracks in isotropic functionally graded materials,"Engineering Fracture Mechanics, vol. 70, no. 1, pp. 1–27, 2003.
88. J. Sladek, V. Sladek, and Ch. Zhang, "Stress analysis in anisotropic functionally graded materials by the MLPG method," Engineering

Analysis with Boundary Elements, vol. 29, no. 6, pp. 597–609, 2005.

89. I. Guiamatsia, B. G. Falzon, G. A. O. Davies, and L. Iannucci, "Element-Free Galerkin modelling of composite damage," Composites Science and Technology, vol. 69, no. 15-16, pp. 2640–2648, 2009.

90. S. S. Ghorashi, S. Mohammadi, and S. R. Sabbagh-Yazdi, "Orthotropic enriched element free Galerkin method for fracture analysis of composites," Engineering Fracture Mechanics, vol. 78, no. 9, pp. 1906–1927, 2011.

91. M. Duflot and H. Nguyen-Dang, "Fatigue crack growth analysis by an enriched meshless method,"Journal of Computational and Applied Mathematics, vol. 168, no. 1-2, pp. 155–164, 2004.

92. U. Andreaus, R. C. Batra, and M. Porfiri, "Vibrations of cracked Euler-Bernoulli beams using meshless local Petrov-Galerkin (MLPG) method," Computer Modeling in Engineering and Sciences, vol. 9, no. 2, pp. 111–131, 2005.

93. S. Wang and H. Liu, "Modeling brittle-ductile failure transition with meshfree method," International Journal of Impact Engineering, vol. 37, no. 7, pp. 783–791, 2010.

94. M. Pant, I. V. Singh, and B. K. Mishra, "Numerical simulation of thermo-elastic fracture problems using element free Galerkin method," International Journal of Mechanical Sciences, vol. 52, no. 12, pp. 1745–1755, 2010.

95. L. Bouhala, A. Makradi, and S. Belouettar, "Thermal and thermo-mechanical influence on crack propagation using an extended mesh free method," Engineering Fracture Mechanics, vol. 88, pp. 35–48, 2012

96. T. Belytschko, Y. Krongauz, M. Fleming, D. Organ, and W. K. S. Liu, "Smoothing and accelerated computations in the element free Galerkin method," Journal of Computational and Applied Mathematics, vol. 74, no. 1-2, pp. 111–126, 1996

97. S. Yoon, C. T. Wu, H. P. Wang, and J. S. Chen, "Efficient meshfree formulation for metal forming simulations," Journal of Engineering Materials and Technology, vol. 123, no. 4, pp. 462–467, 2001.

98. D. Wang and J. S. Chen, "A locking-free meshfree curved beam formulation with the stabilized conforming nodal integration," Computational Mechanics, vol. 39, no. 1, pp. 83–90, 2006

99. A. Khosravifard and M. R. Hematiyan, "A new method for meshless integration in 2D and 3D Galerkin meshfree methods," Engineering Analysis with Boundary Elements, vol. 34, no. 1, pp. 30–40, 2010.
100. H. J. Chung and T. Belytschko, "An error estimate in the EFG method," Computational Mechanics, vol. 21, no. 2, pp. 91–100, 1998.
101. X. Zhuang, C. Heaney, and C. Augarde, "On error control in the element-free Galerkin method,"Engineering Analysis with Boundary Elements, vol. 36, no. 3, pp. 351–360, 2012.
102. H. G. Kim and S. N. Atluri, "Arbitrary placement of secondary nodes, and error control, in the meshless local Petrov-Galerkin (MLPG) method," Computer Modeling in Engineering and Sciences, vol. 1, no. 3, pp. 11–32, 2000.

Chapter 5

Fracture Mechanics of Polymer Mortar Made with Recycled Raw Materials

Marco Antonio Godoy Jurumenha; João Marciano Laredo dos Reis

Theoretical and Applied Mechanics Laboratory - LMTA, Mechanical Engineering Post Graduate Program - PGMEC, Universidade Federal Fluminense - UFF, Rua Passo da Pátria, 156, Bloco E, sala 216, Niteroi, RJ, Brazil

ABSTRACT

The aim of this work is to show that industrial residues could be used in construction applications so that production costs as well as environmental protection can be improved. The fracture properties of polymer mortar manufactured with recycled materials are investigated to evaluate the materials behaviour to crack propagation. The residues used in this work were spent sand from foundry industry as aggregate, unsaturated polyester resin from polyethylene terephthalate (PET) as matrix and polyester textile fibres from garment industry, producing an unique composite material fully from recycled components with low cost. The substitution of fresh by used foundry sand and the insertions of textile fibres contribute to a less brittle behaviour of polymer mortar.

INTRODUCTION

Nowadays, because of the more exigent legislation regarding the environment and the market demand for environmentally friendly products, manufacturers are concerned to develop studies aimed at reducing the environmental impact, through lowering the amount of residues or treating those that are inevitably generated during production processes[1]. High costs associated with raw material extraction, as well as the damage that the extraction causes to the environment, are also important reasons to motivate the use of industrial process residues. Depletion of reliable trustable raw material reserves and conservation of non-renewable sources also contribute to such reuse of waste materials.

The generation of residues is inherent to the casting process, mainly sand from moulds and cores. Therefore, there is growing interest in the re-utilization of this sand, since the amount of residual sand is quite significant. In general, these residues are classified as non-dangerous, class II, according to Brazilian laws. Therefore, recycled foundry sand presents high potential to be used as raw material. For this case, the foundry sand needs no pre-treatment to be used as inert in polymer mortar (PM).

Poly (ethylene terephthalate) (PET) is thermoplastic polyester widely used in applications as diverse as textile fibres, films and moulded products[2]. Among all plastics, PET has received particular attention in terms of post-consumer recycling, due to the relatively large availability of PET bottles from special collection schemes[4].

Numerous ways of recycling disposable beverage bottles have been reported[3], including methods for chemical recycling[3,5], such as methanolsis, glycolsis, hydrolysis, ammonolysis and aminolysis or physical recycling by re-melting[3,4,6-8]. In this case, the recycling of PET bottles was done by glycolysis producing unsaturated polyester resin.

Textiles are manufactured to perform a wide range of functions and are made up of different types of fibres mixed in varying proportions. While the textile industry has a long history of being thrifty with its resources, a large proportion of unnecessary waste is still produced each year, much of which is either incinerated or disposed of in landfill. Textile wastes take many forms and are often complex in nature due to the range of manufacturing specifications required. Complex mixtures of fibres make separation more difficult and more costly, and this have implications for the profitability of textile recycling.

Textile waste arising originates from both the household (consumer) sector and the industrial (manufacturing) sector. Consumer waste generally comprises binned waste or that separated for reuse or recycling, such as unwanted clothing and carpets. Manufacturing waste originates from the processing of raw materials and in the fabrication and production of finished textiles and garments, including cuttings and rejected materials.

The proposal of this work is to use, as inert, recycled foundry sand with organic pollutants in substitution of fresh foundry sand in the manufacturing process of unsaturated polyester resin based on recycled polyethylene terephthalate (PET) polymer mortar. Also, textile fibres from garment industry were used as reinforcement in order to reduce the brittleness of polymer mortar. The fracture characterization of the produced material enables the safe and appropriate use of the material in various precast applications.

MATERIALS AND METHODS

Materials

Polymer mortar formulations were prepared by mixing fresh and recycled foundry sand as aggregates and unsaturated polyester resin recycled from PET as binder. Also textile fibres from garment industry were introduced to the mix proportion by 1 and 2% in weight as reinforcement. Resin content was 12% by weight and no filler was added in formulations. Previous studies carried out by the author[9], considering an extensive experimental program, allowed an optimization of mortar formulations that are now being used in the present work.

The fresh aggregate was green foundry sand with very uniform grains and a mean diameter of 300 μm, produced by JUNDU, design as AG 40-50, meaning aggregate with finesses modulus between 4 and 5. The specific gravity of the fresh sand was 2.63 g/cm³. The foundry sand was previously dried before added to the polymeric resins in an automatic mixer. The recycled sand, like the fresh one, consists primarily of silica sand, coated with a thin film of burnt carbon, residual binder (bentonite, sea coal, resins/chemicals) and dust with a specific gravity of 2.69 g/cm³. It contains cured alkaline-phenolic resin whereas cross-linking is activated with very strong organic acids. The initiator used to promote the free radical to produce polymer mortar due to its high performance, resulting in a high strength and polymerization on the system was a carboxylic acid called Triacetin (glycerol triacetate), which determines the speed of cross-linking. Because of the presence of phenols in foundry sand, there is some concern that precipitation percolating through stockpiles could mobilize leachable fractions, resulting in phenol discharges into surface or ground water supplies.

The polyester resin used, as binder, was unsaturated polyester obtained from recycling PET. Polyester resin is the most used resin to produce polymer mortar due to its high performance, resulting in a high strength and durability against aggressive environments, with low permeability and lower cost when compared to epoxy resins. Polyester resin from PET showed similar results when used as binder to polymer mortar when compared to ordinary polyester resin, with the advantage of low manufacturing cost and processing energy and, of course contributing to reduce the plastic waste[10]. Resin properties are presented in Table 1.

Table 1: Properties of unsaturated polyester resin recycled from PET

Property	Polyester
Viscosity at 25°C μ (cP)	250-350
Density p (g/cm3)	1.09
Heat distortion temperature HDT (°C)	85
Modulus of elasticity E (GPa)	3.3
Flexural strength (MPa)	45
Tensile strength (MPa)	40

Maximum elongation (%)	1

The textile fibres consist of cotton, polyester, silk and rayon. A homogenous single type of textile usually consists of a combination of these materials in various percentages.

The textile cuttings may not be conceived as either an aggregate or reinforcement. It does however contribute to the increase in volume of the mixture (which is the major function of an aggregate) less the weight, and intent to contribute to the increase of polymer mortar ductility textile fibres reinforced polymer mortar were prepared in the same way as plain polymer mortar, with the incorporation of 1 and 2% in weight of chopped textile fibres. The textile waste cuttings are trimmed into average lengths between 2 and 6 cm.

Polymer mortar fracture specimens were compacted in a steel mould of dimensions of 30 × 60 × 250 mm^3 following the specifications of RILEM TC113/PC-2[11]. The specimens were initially cured at room temperature and then post-cured during 3 hours at 80 °C. The samples were notched using a 2 mm diamond saw to a 20 m depth.

Methods

To determine the fracture properties, three-point bending tests were conducted using a universal testing machine with a crosshead speed of 0.5 mm/min. The crack mouth opening displacement (CMOD) was measured using a COD gauge clipped to the bottom of the beam and held in position by two 1.5 mm steel knife edges glued to the specimen, as shown in Figure 1.

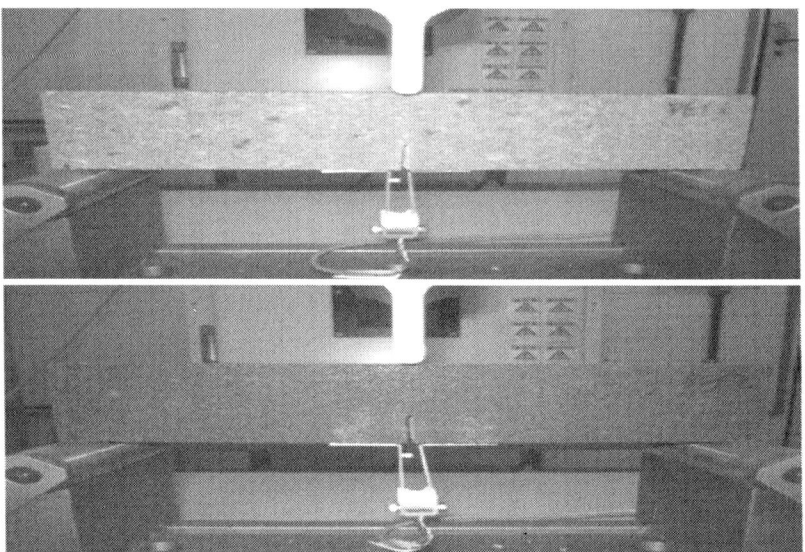

Figure 1: Three-point bending test set-up for polymer mortar with fresh and recycled aggregate.

Fracture toughness, K_{Ic}, and fracture energy, G_f, are the main parameters determined to predict toughness and cumulative energy as crack propagates.

To identify fracture toughness of PC, which is a measurement of a material's resistance to crack extension when the stress state near the crack tip is predominantly plane strain, plastic deformation is limited, and opening mode monotonic load is applied, the Two Parameter Method (TPM)[12] was used. This method is a direct method to calculate fracture parameters and can be expressed as, in (MPa $\sqrt{m}$)[12]

$$K_{Ic} = \frac{3 P_{max} S}{2 W^2 B} \sqrt{\pi \underline{a}} F(\alpha) \tag{1}$$

in which a is the effective critical crack length, is a/W, P_{max} is the measured maximum load [N], S, W and B are the span, depth and width, respectively. The results correspond to the mean values of five tests.

The fracture energy, G_f, was measured according to the RILEM Technical Committee[13], in single edge notched beams, when three-point bending tests are performed. It is the energy necessary to create a unit crack surface and it is also equal to the area defined by softening law. Crack propagation is essentially governed by the mechanical interaction of the aggregates with the polymer matrix.

$$G_f = \frac{W_0 - mg\delta_0}{A_{lig}} \quad (2)$$

where W_0 is the area under the load vs. deflection curve (N/m), m.g is the self-weight of the specimen between supports (kg), δ_0 is the maximum displacement (m), and A_{lig} is the fracture area [d(b-a)] (m²); b and d are the height and width of the beam, respectively[13].

The Young Modulus (E) is calculated from the measured initial compliance C_i using equation

$$E = \frac{6Sa_0V_1(\alpha)}{(C_iW^2B)} \quad (3)$$

in which S is the specimen loading span, a_0 is the initial notch depth, H_0 is the thickness of clip gauge holder, W and B are the beam depth and width respectly according to RILEM[12].

RESULTS AND DISCUSSION

Fracture tests results obtained from 3-point bending tests performed in polymer mortar manufactured with recycled raw materials are presented in Table 2. The specimens were organized in fresh and recycled sand, unreinforced, 1 and 2% textile fibres reinforcement. Polymer mortar manufactured with fresh sand has FS designation and PC formulation with recycled sand has RS designation. The numbers 0, 1 and 2 represent the fibres percentage in the formulations. Five specimens were tested for each formulation and the mean results were plotted.

Table 2: Polymer mortar made with raw materials fracture test results (Avg. ± St.Dev)

Specimens	KIc (MPa $\sqrt{m}$)	Gf (N/m)	E (GPa)
FS0	1.13 ± 0.09	173.24 ± 9.58	16.03 ± 1.68
FS1	0.86 ± 0.16	173.90 ± 6.01	12.39 ± 0.87
FS2	0.53 ± 0.08	162.77 ± 8.36	10.71 ± 0.63
RS0	0.85 ± 0.11	193.48 ± 9.89	10.96 ± 0.78
RS1	0.49 ± 0.04	170.91 ± 7.29	4.31 ±0.67
RS2	0.31 ±0.02	126.12 + 2.29	4.12 ±0.96

It is known from previous studies[9] that textile fibres do not improve polymer mortar mechanical properties. From that knowledge the aim in this research is not improving fracture properties but retard crack propagation turning PM formulations less brittle. Increasing fibre content of fresh sand polymer mortar, the fracture toughness decrease 53.1 and 6% diminish is reported when fracture energy were measured. Also, the modulus of elasticity, E, lowered 33.1% with the increase of textile fibre content.

Figures 2 and 3 displays the average fracture behaviour, both toughness and energy, respectively, of plain and reinforced fresh sand polymer mortar. It is clear from Figures 2 and 3 that increasing fibre content, softer crack propagation is observed. Also, a decrease in the angle between the graph slope and the x-axis demonstrates lower stiffness of fresh sand textile fibre polymer mortar.

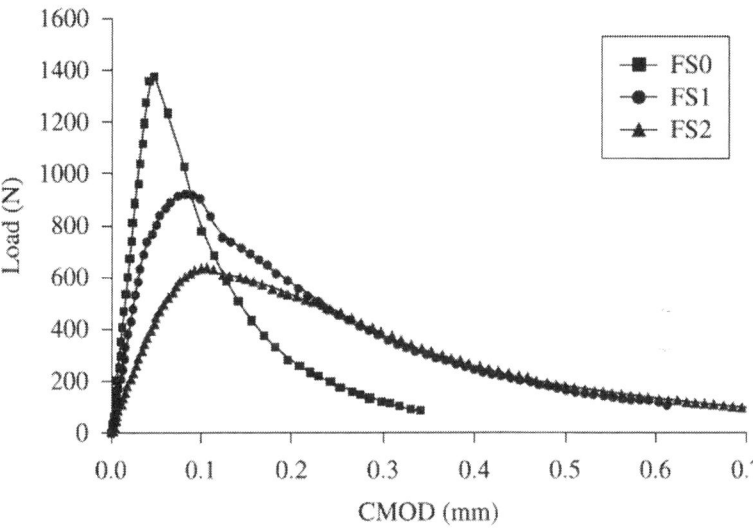

Figure 2: Load vs. CMOD test result of fresh sand textile reinforced polymer mortar.

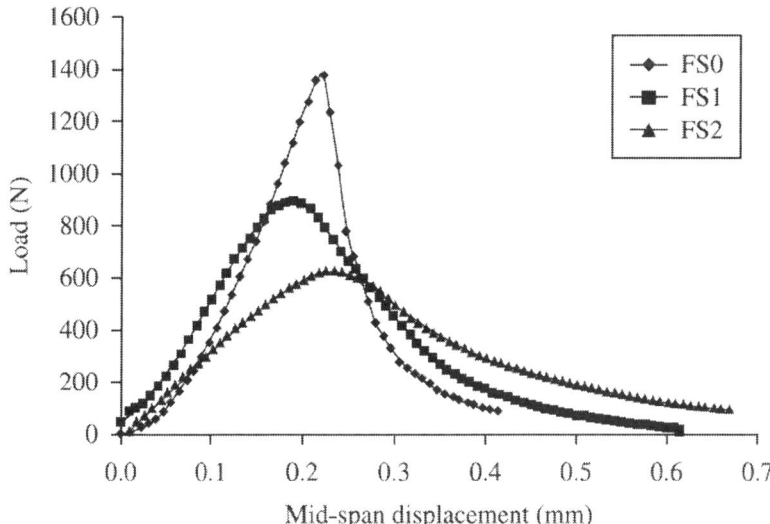

Figure 3: Load vs. Mid-span Displacement of fresh sand textile reinforced polymer mortar.

Following the behaviour reported by fresh sand textile fibre polymer mortar, recycled foundry sand textile fibre polymer mortar also displays decrease in measured parameters such as fracture toughness, energy and modulus elasticity. When textile fibres were added to the mixture, 2% in weight, the fracture toughness decreases 63.5% and fracture energy 34.8%. Substituting the fresh aggregate by a recycled one, the fracture properties of unsaturated polyester mortar are deeply affected. Fresh sand unreinforced polymer mortar is tougher and stiffer than recycled unreinforced polymer mortar but polymer mortar manufactured with recycled aggregate can retain more energy during crack propagation than fresh sand polymer mortar. When textile fibres are inserted in the mixture as fillers or reinforcement the fresh sand polymer mortar retains more energy than recycled ones.

Figures 4 and 5 plot the average evolution of recycled sand textile polymer mortar during the three-point bending tests. Again, it is clear from the graphs that the increment of fibre content contributes to lowering brittleness and fracture properties polymer mortar, both fresh and recycled sand.

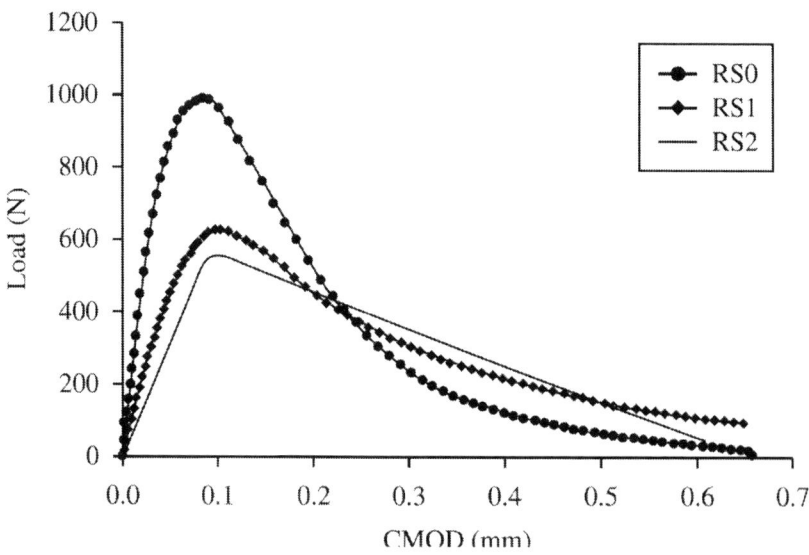

Figure 4: Load vs. CMOD test result of recycled sand textile reinforced polymer mortar.

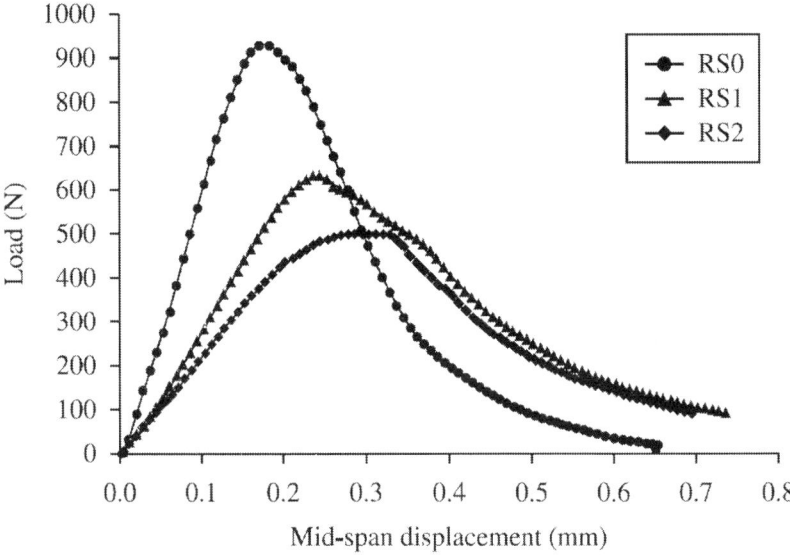

Figure 5: Load vs. Mid-span Displacement of recycled sand textile reinforced polymer mortar.

A comparative fracture toughness chart is presented in Figure 6. Fresh sand polymer mortar reinforced with 1% of textile fibres behaves similar to unreinforced recycled sand polymer mortar. Their resistance to crack propagation are at the same level. The same performance is observed for 2% textile fresh sand polymer mortar compared to 1% textile fibre recycled sand polymer mortar. The brittleness of both, fresh and recycled sand, polymer mortar decreases when textile fibres are added to the mixture.

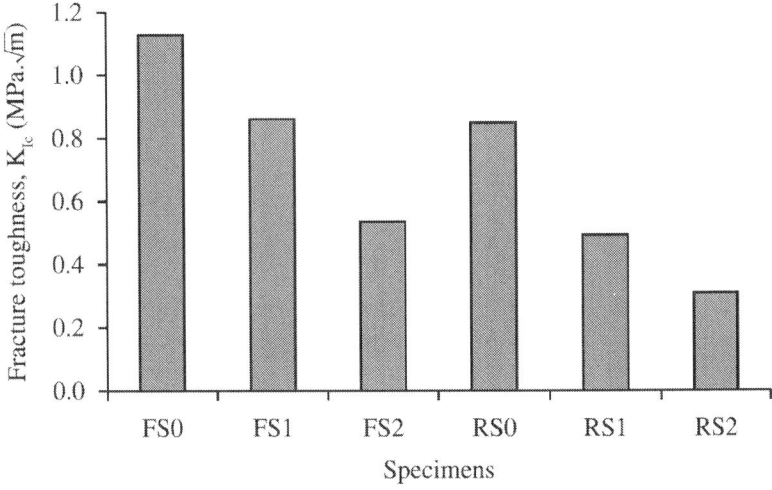

Figure 6: Polymer mortar fracture toughness comparison.

According to Figure 7, instead of brittleness increase, the energy during the fracture process of all formulations is at the same level. The insertion of textile fibres in fresh polymer mortar does not prevent the initiation of crack, but when crack occurs the propagation is slower due to fibre bridging between textile fibres and the polymer matrix.

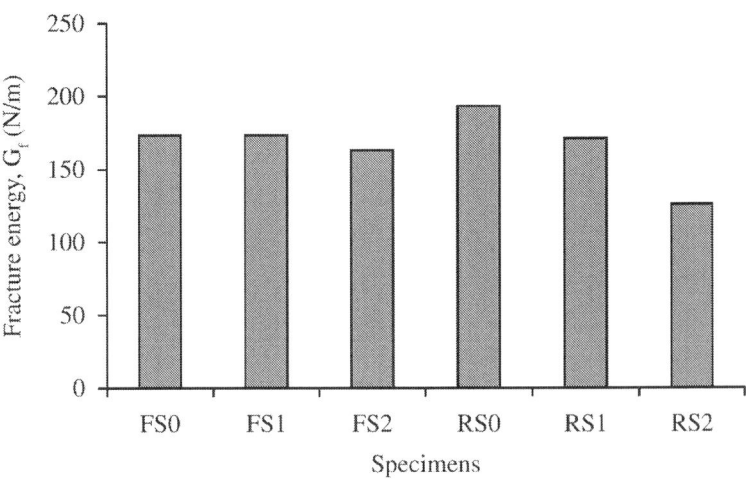

Figure 7: Polymer mortar fracture energy comparison.

CONCLUSIONS

The fracture properties of fresh and recycled sand polymer mortar with unsaturated polyester resin from recycled PET, as binder, were investigated in this research work. Also, textile fibres from garment industry were added to the mixture producing a unique material complete manufactured with waste as raw materials, recycled sand, textile fibres and recycled unsaturated polyester resin from PET.

The substitution of fresh by used foundry sand contributes to decrease crack resistance and propagation. Again, the insertions of textile fibres diminish the fracture toughness becoming polymer mortar less resistant to crack propagation and also changes in the post-peak status are reported. Recycled foundry sand and waste textile fibres from garment industry could be very conveniently used in making good quality polymer mortar and construction materials.

ACKNOWLEDGEMENTS

The financial support of Rio de Janeiro State Funding, FAPERJ, and Research and Teaching National Council, CNPq, are gratefully and acknowledged.

REFERENCES

1. Bragança SR, Vicenzi J, Guerino K and Bergmann CP. Recycling of iron foundry sand and glass waste as raw material for production of whiteware. *Waste Management Research*. 2006; 24:60-6.
2. Gupta VB and Bashir Z. PET fibers, films, and bottles. In: Fakirov S, editor. *Handbook of thermoplastic polyesters*. Germany: Wiley-VCH; 2002. p. 317-388.
3. Nadkarni VM. Recycling of polyesters. In: Fakirov S, editor. *Handbook of thermoplastic polyesters*. Germany: Wiley-VCH; 2002, p. 1223-49.
4. Paci M and La Mantia FP. Competition between degradation and chain extension during processing of reclaimed polyethylene terephthalate. *Polymer Degradation and Stability*. 1998; 61:417-20.

5. Spychaj T. Chemical recycling of PET: methods and products. In: Fakirov S, editor. *Handbook of thermoplastic polyesters.* Germany: Wiley-VCH; 2002, p.1251-90.
6. Giannotta G, Po R, Cardi N, Tampellini E, Occhiello E, Garbassi F et al. Processing effects on poly(ethylene-terephthalate) from bottle scraps. *Polymer Engineering Science* 1994; 34:1219-23.
7. Paci M and La Mantia FP. Influence of small amounts of polyvinylchloride on the recycling of polyethylene terephthalate. *Polymer Degradation and Stability* 1999; 63:11-14.
8. Frounchi M. Studies on degradation of PET in mechanical recycling. *Macromolecular Symposia.* 1999; 144:465-9.
9. Reis JML. Effect of textile waste on the mechanical properties of polymer concrete. *Materials Research.* 2009; 12:63-67.
10. Rebeiz KS. Time-temperature properties of polymer concrete using recycled pet. *Cement & Concrete Composites.* 1995; 17:119-124.
11. RILEM. *PC-2: Method of making polymer concrete and mortar specimens.* Technical committee TC-113. Test methods for concrete-polymer composites (CPT). International union of testing and research laboratories for materials and structures; 1995.
12. Jenq YS and Shah SP. Two parameter fracture model for concrete. *Journal Engineering Mechanics.* 1985; 111:1227-41.
13. RILEM. 50-FMC. Determination of fracture energy of mortar and concrete by means of three-point bend test on notched beams. *Materials Structures.* 1985; 18:285-90.

Chapter 6

In situ Experimental Mechanics of Nanomaterials at the Atomic Scale

Lihua Wang[1], Ze Zhang[1,2], and Xiaodong Han[1]

[1]Institute of Microstructure and Properties of Advanced Materials, Beijing University of Technology, Beijing, China
[2]Department of Materials Science and Engineering and State Key Laboratory of Silicon Materials, Hangzhou, China

ABSTRACT

Sub-micron and nanostructured materials exhibit high strength, ultra-large elasticity and unusual plastic deformation behaviors. These properties are important for their applications as building blocks for the fabrication of nano- and micro-devices as well as for their use as components for composite materials, high-strength structural and novel functional materials. These nano-related deformation and mechanical behaviors, which are derived from possible size and dimensional effects and the low density of defects, are considerably different from their conventional bulk counterparts. The atomic-scale understanding of the microstructural evolution process of nanomaterials when they are subjected to external stress is crucial for understanding these 'unusual' phenomena and is important for designing new materials, novel structures and applications. This review presents the recent developments in the methods, techniques, instrumentation and scientific progress for atomic-scale *in situ* deformation dynamics on nanomaterials, including nanowires, nanotubes, nanocrystals, nanofilms and polycrystalline nanomaterials. The unusual dislocation initiation, partial-full dislocation transition, crystalline–amorphous transitions and fracture phenomena related to the experimental mechanics of the nanomaterials are reviewed. Current limitations and future aspects using *in situ* high-resolution transmission electron microscopy of nanomaterials are also discussed. A new research field of *in situ* experimental mechanics at the atomic scale is thus expected.

INTRODUCTION

Recent studies on nanostructured materials, including nanowires (NWs),[1, 2] nanotubes (NTs),[1, 3] nanocrystals (NC),[4] micro/nano-pillars (NPs)[5, 6, 7] and nanocrystalline,[8, 9] have revealed a variety of 'unusual deformation' phenomena compared with their bulk counterparts, such as high strength, nano-piezoelectric effects and unusual plastic deformation behaviors. Nanostructured materials can apparently sustain a larger dynamic range of elastic and plastic strains than conventional materials. The results from these studies indicate that the fundamental dislocation processes that initiate and sustain plastic flow and fracture in nanoscale materials are considerably different than in their conventional bulk counterparts. These 'unusual' phenomena not only allow these materials to possess excellent mechanical properties but also enable the tuning of their band structures and related novel electronic, magnetic, optical, photonic and catalytic properties. Revealing the atomic-scale deformation mechanisms of nanomaterials (NM) and controlling their elastic and plastic properties are useful for realizing the desired mechanical, physical and chemical properties through the application of stress or strain.

Although extensive studies have been conducted to investigate the mechanical properties of NM,[10] the majority of the atomic mechanisms are based on computer simulations, which may suffer from inaccuracies due to the empirical or semi-empirical interatomic potentials, grain boundary (GB) structures and high strain rates.[11,12] This article focuses on recent *in situ* atomic-scale experimental studies on the deformation behaviors of NM. We briefly introduce some important techniques and methods that have been used for the time-resolved visualization of nano-mechanics that utilize *in situ* microscopy, specifically, the techniques used for gaining an atomic-scale understanding of the deformation behaviors of NM. The size effects that lead to the ultra-large elasticity in these materials will be discussed. The atomic-scale *in situ* transmission electron microscopic (TEM) investigations on the elastic–plastic transition and the plastic deformation mechanisms of NM will be reviewed.

EXPERIMENTAL TECHNIQUES BY IN SITU MICROSCOPY

The early testing devices for NWs and NTs were based on the atomic force microscopy (AFM) technique.[1,2,3] These devices enabled the direct determination of force as a function of displacement and revealed the unusual mechanical properties of NM.[13,14,15] However, these techniques cannot normally reveal the actual deformation mechanisms that involve dislocation activities in the NM because of the difficulties in interpreting the displacement–force correlations to details of the dislocation initiation and interaction activities.[16]

TEM is one of the most powerful and effective techniques that has nanoscale and atomic-level resolution capabilities along with the ability to obtain crystallographic and chemical information. *In situ* TEM experiments have been used since the 1960s.[17] During the past decade, the use of TEM for investigating *in situ* mechanics and the related physics of materials has been one of the most interesting research fields due to the unexpected and unusual mechanical and physical size-effects in materials with micron/nano dimensions or volumes.[4,5,6,7,9,10,13,14,15,16,17,18]

The simplest approach for performing *in situ* deformation experiments in a TEM utilizes a conventional TEM single-tilt straining stage, which can tilt the strained/stressed sample along one axis. With this stage, the total elongation of the sample can be determined by monitoring the elongation from subsequent TEM images. In most cases, the cross-sectional area of the sample being strained is unknown. These aspects prevent a correct stress measurement. Using the Gatan straining holder, the microstructural evolution of nanocrystalline thin films and single crystals[19, 20, 21] have been recorded *in situ* during the deformation process in the TEM.

Quantifying the stress–strain relationship is an important requirement in regular mechanical property studies. Based on nanoindentation techniques, AFM and scanning probe microscopic tips have been developed that can be used in a TEM.[22] The samples are normally mounted on a piezo-driven support within the stage. Load sensors with high resolution are available for different regions of interest. The displacement is either measured capacitively or is deduced from the applied piezo-motion. The NanoFactory TEM–STM (scanning tunneling microscopy)/AFM holders[22] and the Hysitron picoindenter[6, 7] are the most commonly used devices for deformation testing with the ability to quantify the stress–strain relationships. With stress versus strain curves, the TEM–STM holder could also apply a specific bias and measure the current response of the NMs. Several other in-lab-developed techniques and devices have also been used in studies of the deformation mechanisms and strained physics of NM.[23, 24, 25, 26, 27, 28, 29]

The micro-electromechanical system (MEMS) is another important approach for the *in situ* microscopy mechanics by TEM.[8, 30] These devices are based on Si technology, and film deposition, lithography and etching techniques are used to design actuators and force sensors on a chip. One of the first successful MEMS devices for *in situ* TEM deformation studies of metallic lines was developed by Saif's group.[30] The sample was co-fabricated with the MEMS structure and suspended as a line on the device. During loading, the gap distance between the sensing beams changes proportionally to their stiffness, while the stiffness of the sensing beams is measured using the nanoindentation technique.

The aforementioned commercial stages or MEMS devices can provide quantitative stress–strain data along with revealing the microstructural evolution process of the strained materials. Numerous investigations[31, 32, 33, 34] have provided valuable quantitative insight into the dislocation mechanisms of NM using *in situ* TEM observations. However, these devices only have single-tilt capabilities, which makes reliable investigations of linear and planar defects inconvenient because a 'double-tilt' ability is generally required to obtain the desired 'two-beam' condition or the ideal crystallographic orientation. Direct atomic-scale investigations using these devices are difficult. Therefore, for reliable defect investigations in crystalline materials, developing microscopy mechanic devices with double-tilt capabilities is necessary.

With the use of a conventional double-tilting stage and qualitative irradiation of multiwalled carbon nanotubes (MWCNTs),[27] ultra-high pressures can accumulate in the cores of the MWCNTs. With this method, Sun et al.[27, 35] investigated the deformation dynamics of NCs through the controlled irradiation of MWCNTs. Han et al.[25, 26, 36, 37, 38] developed an alternative TEM grid technique that can bend or conduct axial tensile experiments on individual NWs using a pre-broken colloidal thin film on the TEM grid. The above two methods do not require mechanical tensile attachments, and the specimen could be tilted along a pair of two orthogonal directions with large angles. Therefore, with a proper observing direction, the deformation process in individual NWs or NCs can be recorded *in situ* at the atomic scale. However, the strain rate and deformation mode is normally uncontrollable.

Controllable *in situ* straining experiments without sacrificing the double-tilt capability in the TEM was developed by exploiting the differences in the thermal expansion coefficients between different materials. This method has been used to investigate plasticity in thin films[39, 40] on substrates, and it requires a conventional heating stage. The mismatch in the thermal expansion coefficients between the film and the substrate will cause the film to experience tensile or compressive stress. With this method, the strain applied by the substrate on the thin film is very small, and the film–substrate diffusion or chemical reactions during heating will complicate the interpretation of the results.

Based on the thermal bimetallic technique that has been used in SEM analyses,[28] Han et al.[38] developed a novel *in situ* controllable tensile testing device for TEM measurements, which can slowly and gently deform the NWs, NTs, NPs and nanocrystalline thin films. Their method can even measure regular TEM samples assisted by focused ion-beam fabrication while retaining the double-tilt capability for performing high-resolution TEM (HRTEM) observations and regular 'two-beam' dark field imaging investigations.[29] The strain rate is also controllable in the range from 10^{-2} to 10^{-5} s^{-1}. The TEM extensor is composed of two thermally actuated bimetallic strips. The conventional TEM heating stage, now with the fabricated bimetallic strips, can therefore serve as a double-tilt, displacement-controlled tensile stage. With this device, investigations of the plastic deformation behavior of nanocrystalline materials,[29,41,42] NCs[43,44,45] and metallic glasses[46,47] can be conducted and observed *in situ* at the atomic scale.

SIZE EFFECT OF ULTRA-LARGE ELASTICITY IN NWS AND NCS

It has been hypothesized that the ideal elastic strain of metallic crystals is on the order of 8%.[10] Small samples always sustain ultra-high elastic strain before yielding.[2,4,5,6,7,8] This behavior is in contrast to conventional bulk metals, which can only deform on the order of 0.2% elastic strain. The majority of the applied strain of bulk metallic materials is performed by inelastic strain, and the strength of the material is primarily increased through work hardening.

The ultra-large elasticity in small-sized samples was discovered as early as 1924 by G. F. Taylor,[48] who observed that antimony wires with a diameter of 30 μm can be repeatedly bent without breaking, although bulk antimony is very brittle. In 1949, Bragg and Lomer[49] reported an experimental investigation on the *in situ* deformation of an extended raft of bubbles floating on a soap solution and observed that the bubble crystal can be elastically deformed to approximately 10%. Later, Herring and Galt[50] observed that tin whiskers with a diameter of 1.8 μm can sustain 2–3% elastic strain, though bulk tin can only sustain approximately 0.01% elastic strain. Subsequent mechanical tests performed on a variety of whiskers,[10,39] NWs,[5,6] NC[51] and NTs[1,3] have revealed that these micro- and nanoscale components can, in fact, sustain large elastic strains. In the past 10 years, a wide range of metallic micro- and nanoscale pillars have been used to investigate the deformation mechanism of small-sized samples.[52, 53, 54, 55] The majority of the results convincingly indicate that the yield and flow stresses increase when the sizes of the samples decrease, which is a phenomenon known as 'smaller is stronger.' This remarkable trend has been comprehensively reviewed by Zhu and Li,[10] Dehm,[15, 39] Uchic et al.,[52] Kraft et al.,[53] and Greer et al.[55]

For bulk materials, we normally only measure the yield strength for the motion of pre-existing dislocations. Pristine crystals of the micro- and nanoscale samples can be produced to minimize the number of defects that inevitably exist in their bulk counterparts. This procedure offers the opportunity to observe large elastic strains close to the theoretical limit.[10] This prediction was directly observed by Yue et al.[43] Using the bimetallic tensile technique, the process of lengthening the atomic bonds was captured and an elastic strain that approached the theoretical elastic limit was observed in Cu NWs. During the deformation of a Cu NW with a diameter of approximately 5.8 nm along the [001] direction, the lattice experienced an approximate elastic strain of 7.2% along the [001] direction, which was directly observed. After the NW was fractured, the lattice returned to its initial value.

DIRECT ATOMIC MECHANISMS OF THE SIZE EFFECTS ON THE UNUSUAL PLASTICITY OF SINGLE CRYSTALLINE FCC (FACE CENTER CUBIC) METALS

The ultra-high strength achieved in NCs implies that their deformation behaviors are considerably different from their bulk counterparts. The uniaxial compression methodology was first introduced by Uchic et al.[5] Greer et al.[56] extended this method to the nanoscale region; they observed that single crystalline Au NPs exhibited unprecedented strengths that were nearly 50 times greater than their bulk counterparts. More recently, the understanding of elasticity and plasticity in small volumes has been further enriched through tensile experiments performed by Mompioua et al.,[32] Kiener et al.,[57] and Dehm et al.[58] Because the deformation mechanism of these micro- NPs have been comprehensively reviewed by Zhu and Li,[10] Dehm,[15, 39, 58] Uchic et al.,[52] Kraft et al.,[53] Legros et al.,[59] and Greer et al.,[55] we only provide a brief summary here. Next, we focus on the atomic-scale *in situ* TEM investigations on NMs that are <100 nm.

Several theories have been proposed to explain this size-dependent strengthening behavior. Two prominent deformation mechanisms have been proposed, which include dislocation starvation[60] and the single-arm source theory.[61] In the former, the plasticity results from the surface nucleation of dislocations once all the pre-existing mobile dislocations have been annihilated at the free pillar surface, and this process was observed *in situ* by Minor and colleagues[6, 33] and Kiener and Minor.[7] In the latter, the creation of dislocations is caused by the operation of single-arm sources; the increase in strength arises from the progressively harder operation of dislocation sources with a reduction in the pillar diameter. These concepts contrast the classical plasticity model, in which the dislocations multiply, and therefore, the overall dislocation density increases, which results in work hardening. Because the diameters of FIB-fabricated NPs are >100 nm, obtaining atomic-scale images of these NPs is difficult. Therefore, the atomic-scale deformation mechanisms of these NPs are highly reliant on molecular dynamics (MD) simulations.

As the size of the NCs continues to decrease (<100 nm), it appears that the size effect has an obvious effect on the types of dislocations, which changed from full dislocations (d~>200 nm) to partial dislocations. Seo et al.[62] reported that defect-free Au NWs (d~100 nm) exhibit superplasticity upon tensile deformation, which is associated with twin boundaries and the propagation of partial dislocations. Sedlmayr et al.[63] observed partial/twinning-mediated plasticity in Au nanowhiskers with diameters between 40 to 200 nm. Partial dislocation emission was also observed in the sub-20-nm-sized gold NWs.[64] Yue et al.[44] quantitatively revealed an obvious effect of the sample dimensions on the plasticity mechanisms using *in situ* tensile tests of Cu single crystalline NWs with diameters between 1000 to 70 nm in a HRTEM. When the size of the single-crystal NW size was reduced to <~150 nm, the normal full dislocation slip was overwhelmed by partial dislocation-mediated plasticity. In fact, the impact of size-dependent dislocation mechanisms has also been observed in thin film.[65]

For NWs <20 nm, the AFM/STM tip-based technique is one of the most powerful approaches for probing the deformation properties. By dipping the STM (TEM–STM holder) probe tips into Au, Au NWs with diameters of a few nanometers can be obtained because of its high adherence ability. This method was first used by Agraite et al.[22] Following this report, the deformation behaviors of Au NWs with various diameters (even <10 nm) have been investigated. Figure 1 presents a typical *in situ* atomic-scale observation, in which the partial dislocations emitted from free surfaces dominated the deformation of the d~10-nm-sized gold NWs.[66] Figures 1a–c present sequential HRTEM images that illustrate the entire dislocation process, including the nucleation of a leading partial dislocation from a surface step (Figure 1a), the stacking fault (SF; as indicated by an arrow in the inset of Figure 1b) and the trailing partial dislocation, which eliminates the SF (Figure 1c).

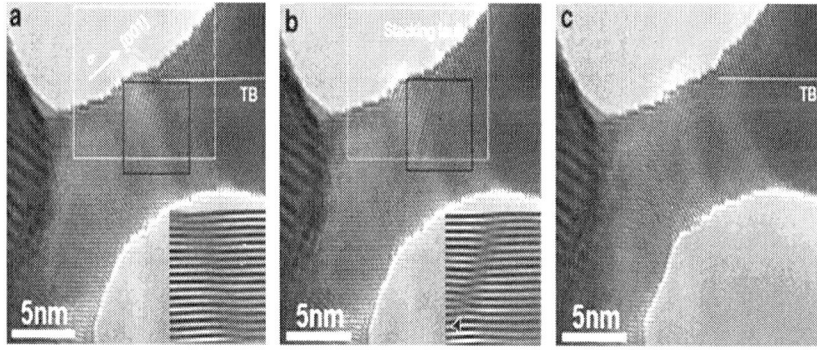

Figure 1: Sequential high-resolution transmission electron microscopic images revealing the emission of a dislocation from a free surface. (a) No dislocations were observed. (b) A leading partial dislocation emission and resulting stacking faults were observed. (c) A trailing partial emission eliminates the stacking fault.

When further reducing the size of the NC to approximately 6 nm, the *in situ* atomic-scale observation indicates that lattice slips were the dominant plastic events.[67] For the 3-nm-sized NC, Kizuka[68] observed that only lattice slip occurs. The lattice slip on the {111} planes was also observed by Matsuda and Kizuka[69] in Pd NWs. The fcc–bcc phase transition was observed *in situ* and in real-time in a d~1.8-nm-sized Au junction.[70]

In 1998, Ohnishi et al.[71] reported a suspended, single chain of gold atoms. No dislocations, lattice sliding or phase transitions were observed in this small-sized NW of approximately 1 nm. The single atom chain was formed by breaking the atomic bonds one by one. Figures 2a–f present TEM images of approximately 1 nm gold NWs formed between the substrate and the tip during the withdrawal of the tip. The dark lines in the gold NW represent rows of gold atoms that span the distance from the substrate to the tip. Figure 2g presents typical HRTEM images of a single chain of gold atoms. Bettini et al.[72] observed the *in situ* formation of single atomic chains of Ag and the Au–Ag alloy.

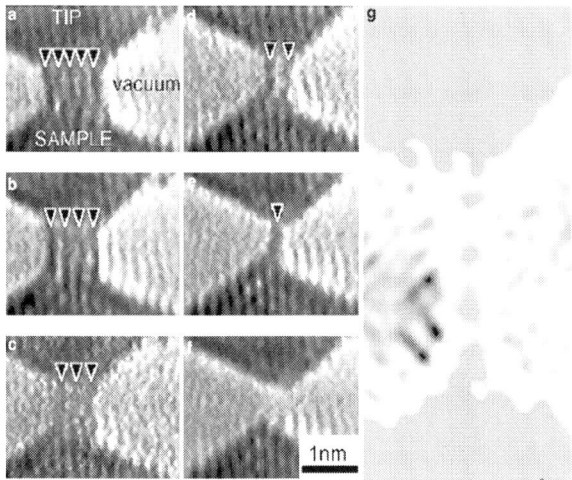

Figure 2: A gold bridge formed between the gold tip (top) and gold substrate (bottom), thinned from (a–e) and ruptured at (f). Dark lines indicated by arrows are rows of gold atoms. (g) transmission electron microscopic image of a single chain of atoms. Figure from Ohnishi et al.[71]

It is clear that the deformation mechanisms change as the crystal size decreases. As shown in Figure 3, when the diameter of the NC is greater than approximately 200 nm, full dislocations dominate the plasticity, whereas for smaller NCs, partial dislocations are prevalent (200–10 nm). As the diameter continued to decrease, new deformation phenomena, such as lattice slip on the {111} planes, phase transitions and fracturing of the bonding atoms, were observed.

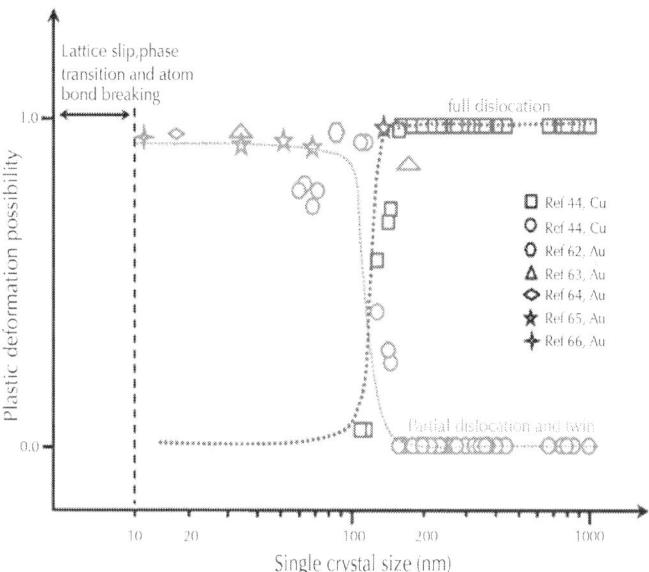

Figure 3: Sample size dependence of two plasticity mechanisms: relative contribution to the overall plastic strain experienced by the sample region under observation, from full dislocation slip (blue symbols) versus partial dislocation (red symbols)-mediated processes.

DIRECT ATOMIC MECHANISMS OF THE SIZE EFFECT ON THE UNUSUAL PLASTICITY IN SEMICONDUCTOR NWS

Bulk semiconducting and ceramic materials are normally brittle and fracture upon any mechanical deformation for shape changes at room temperature (RT). However, when the size of the material decreases to a small scale, the defect-free structure normally makes the NM survive at high fracture stresses,[10, 25, 26, 51, 73] which could eventually provide the materials with the ability to overcome the critical resolved shear stresses and nucleate ductile dislocations or make these ductile-featured dislocations mobile. The fracture and deformation behaviors of NMs can be significantly different from that of their bulk counterparts. In 2007, Han et al.[26] directly observed the unusually large bending strain plasticity in ceramic SiC NWs close to RT. The approximate bending strain was calculated using the traditional formula $\varepsilon_{strain}=r/(r+R)\%$,[74] where R is the bending curvature and r is the radius of the Si NW. In these studies, the bending strain in the figures or text represented the maximum strain in the NW. *In situ* atomic-scale observations revealed that the plasticity of the SiC NWs was accompanied by a process of increased dislocation density during the early stages, followed by an obvious lattice distortion and then amorphization in the most strained region of the NW. Using the thermal bimetallic technique,[28] superplastic elongation ability was observed in the SiC NWs. This RT plasticity was also observed in Si NWs by conducting axial tensile[25] and bending tests[36] in the TEM. *In situ* atomic-scale TEM images revealed a considerable density of dislocations, and a crystalline–amorphous (c–a) transition is responsible for the ultra-large plasticity character of the Si NW. Following this report, large plasticity and the c–a transition have also been observed in other semiconductor NWs. Smith et al.[75] observed that Ge NWs become amorphous at the point of maximum strain of 17%. Asthana et al.[76] observed the c–a transition in the highly compressed region of the [0001] ZnO NWs after a number of loading and unloading cycles. Most recently, Tang et al.[77] revealed that under tension, Si NWs elastically deformed until an abrupt brittle fracture. Under a larger bending strain of >20%, plastic deformation occurred because of dislocations.

Uniaxial compression tests were performed on Si pillars in the size range of 1–200 nm.[78] When the diameter of the pillar falls below a critical value (between 310 and 400 nm), it exhibits ductility. The brittle-ductile transition in GaAs has also been observed at RT.[79] *In situ* TEM compressive experiments on the Si NWs and particles have also been conducted.[80, 81, 82] Highly ductile features, plastic dislocations and strong strain hardening were directly observed. These results apparently indicated that for the semiconducting or ceramic materials, the RT brittle–ductile transition can be realized by reducing the size of the materials.

Atomic-scale imaging is an important technique for revealing localized or incipient phase transition phenomena in the deformation dynamics of crystalline materials. Wang et al.,[83] for the first time, directly observed an atomic mechanism for the crystalline–amorphous (c–a) transition through a dislocation reaction in ultra-large strained (up to 14%) Si NWs. The direct dynamic atomic-scale observations revealed that partial and full dislocation nucleation, motion and interaction and the c–a transition were responsible for absorbing the ultra-large strain during the bending of the Si NWs. Full dislocations were nucleated and then formed a Lomer dislocation by reaction. The continuous straining on the Lomer dislocations induced a c–a transition in the Si NWs. These results provide a direct explanation for the ultra-large straining ability and the c–a transition mechanism for those semiconductor and ceramic nanostructures.[25, 26, 36, 75, 76, 77]

DIRECT ATOMIC MECHANISMS OF PLASTIC DEFORMATION IN NANOCRYSTALLINE MATERIALS

The plastic deformation behaviors of bulk polycrystalline metals are well understood.[84] However, the precise nature of the plastic deformation mechanism in nanocrystalline materials is still not fully certain. As proposed by MD simulations, when the grain size (diameter d) is less than approximately 15 nm, the dislocation activities subside, which may completely give way to the GB-mediated plasticity.[11,12] This phenomenon is often referred to as the inverse Hall–Petch effect. However, the existence of this inverse Hall-Petch effect is still under debate.[85]

For example, many previous *in situ* TEM studies have observed GB-mediated plasticity. In 1995, Ke et al.[20] observed that grain rotation is the dominant plastic deformation mechanism based on *in situ* HRTEM measurements. Shan et al.[86] *in situ* observed grain rotation in nanocrystalline Ni using dark-field TEM images. The GB migration in Al was observed[31, 87] based on the contrast changes in the TEM images. However, there are also many TEM observations that contrast with the above MD simulations and TEM observations, including *in situ* ones, which suggest that the dislocations are highly active even in the approximately 10-nm-sized grains.[29, 41, 42, 88]

Recently, Wang et al.[41] developed a new bimetallic technique that can perform *in situ* axial tensile deformation on NM at the atomic scale

(Figure 4a). With this method, the direct atomic-scale observation revealed that an array of inter-grain dislocations can also induce grain rotation in nanocrystalline gold.[42] A high activity of dislocation behaviors were also observed in nanocrystalline Pt with grain sizes less than approximately 10 nm.[29, 41] As shown in Figure 4b, a Lomer dislocation was observed in the grain, which was formed by dislocation reactions. With further loading, the destruction and the reformation of the Lomer lock in the grain was also observed, as shown in Figures 4c and d. In addition to the formation of Lomer dislocations, the dislocation annihilation and storage of full dislocations in the d~10 nm were also detected.[41] These dislocation activities were also observed by HRTEM examinations in Ni,[89] Al[90] and Cu nanocrystalline.

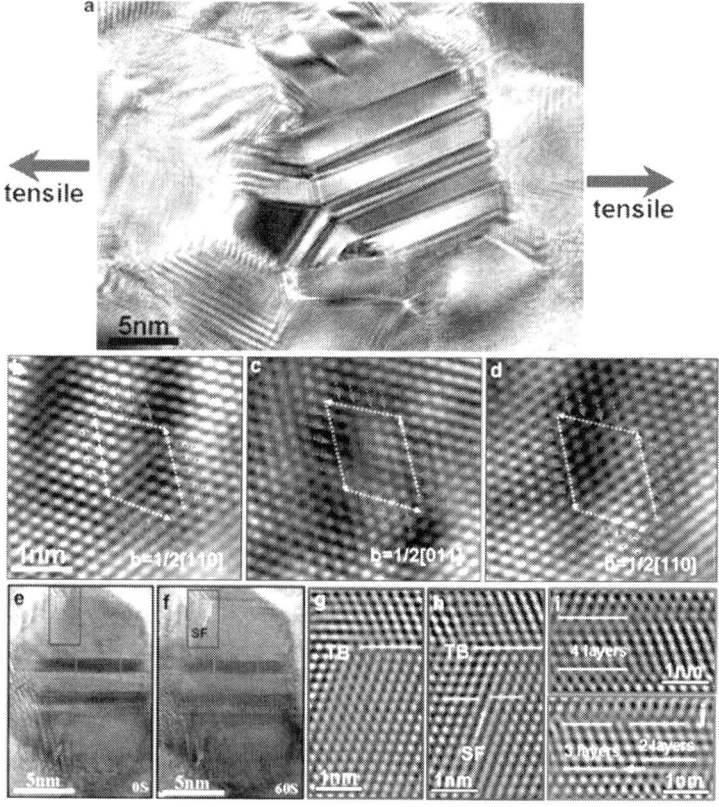

Figure 4: (a) A schematic illustration of the dynamic processes of the nanocrystal can be recorded *in situ* during tensile loading with the bimetallic tech-

niques. (b–d) Enlarged high-resolution transmission electron microscopic (HRTEM) image of the same region that was recorded at a different stage of deformation. (b) Lomer dislocations are observed in the grain. (c) Under further loading, the Lomer dislocations undergo a destruction process; only 60° of full dislocations were observed. (d) We further discovered the re-formation of Lomer locks in the grain. (e, f) Two consecutive HRTEM images. (g, h) Partial dislocation nucleated and glided incline toward the twin boundary (TB) in the thicker twin. (i,j) Partial dislocation nucleated and glided incline parallel to the TB in the thinner twin. Figure from Wang et al.[29] Copyright 2010 The American Physical Society. SF, stacking fault.

It is still uncertain whether the inverse Hall–Petch effect really exists.[85] However, a considerable number of *in situ* and *ex situ* TEM observations exhibit the same trend: the grain size has an obvious effect on the type of dislocations and the deformation twins. Wang et al.[29] provided direct experimental evidence for the transition from full dislocation to partial dislocation with decreasing *grain size*. For grains larger than d~10 nm, full dislocation activities are the dominant deformation mode. For the smaller grains (d~<10 nm), partial dislocations that generate SFs are prevalent. Li et al.[89] presented an *in situ* atomic-scale observation of reversible SFs in small-sized nanocrystalline Ni and deformation twinning in Ni and Cu was also observed *in situ* at the atomic scale.[91] Partial dislocation or deformation twinning have also frequently been observed in other small-sized FCC metals by post-mortem examination,[29, 89, 90, 91, 92] even for those metals with high SFs or twinning energies, such as Al,[90, 92] Pt[29] and Ni.[89, 91, 92] This size effect on the dislocation type is very similar to that observed for single crystals. Figure 5 summarizes the size-dependent plastic deformation mechanisms of Pt nanocrystalline samples; for grain sizes above a critical value, full dislocations will dominate the plastic deformation, whereas below this critical value, partial dislocations will gradually control the plastic deformation. As the grain size continues to decrease, it will transition to GB-mediated plastic deformation (theoretical inverse Hall–Petch effect), although this issue remains uncertain. For different materials, the transition region is different.

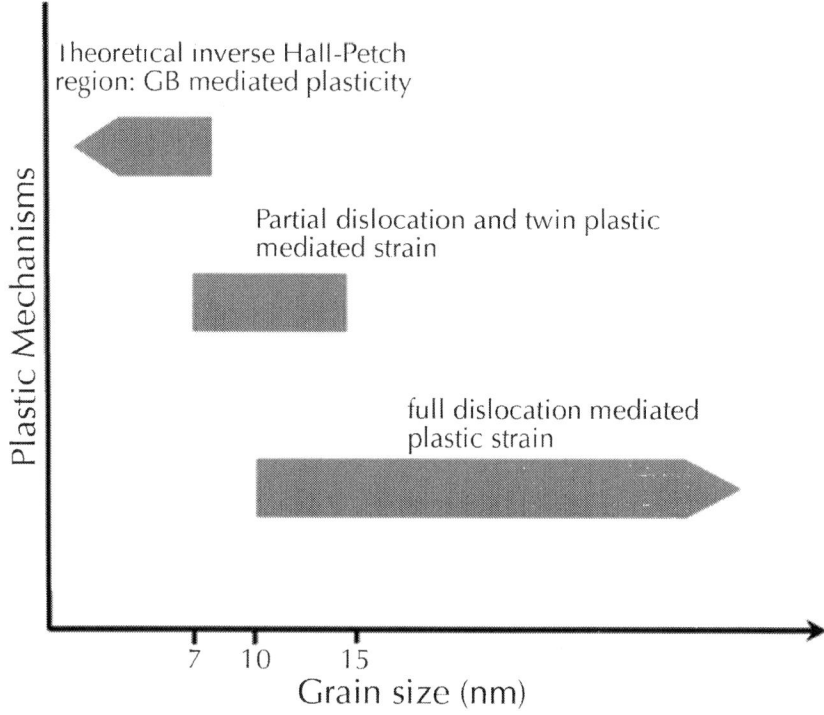

Figure 5: Grain size effect on the plastic deformation mechanisms in Pt nanocrystalline samples. As the grain size decreases, there is a transition from full dislocation to the partial dislocation. When the grain size is less than approximately 7 nm, grain boundary (GB) will mediate the plastic deformation.

Twin-structured crystalline materials always exhibit high strength and high ductility.[93] For twin-structure crystalline copper (with submicron-sized grains), Lu et al.[93] revealed that the strength increases with decreasing twin thickness, which reaches a maximum at 15 nm followed by a softening at smaller values. A strong twin thickness effect on the dislocation behaviors was also proposed based on post-mortem observation: dislocation–dislocation interaction hardening in coarse twins, and dislocation–twin boundary (TB) interaction hardening in fine twins. This claim was confirmed by Wang et al.[94] The authors' *in situ* atomic-scale observations revealed that the extended dislocations can hardly cross the TB in Cu with growth twins. Wang[95] and Li[19] investigated the stability of growth twins in Cu using *in situ* nanoindentation.

For twin-structured nanocrystalline (grain size <100 nm) materials, the understanding of the atomic mechanism of deformation dynamics is primarily based on MD simulations. Li et al.[96] reported an 'inverse Hall–Petch' effect on the thickness of a twin plate. There is a critical twin thickness (c) for a given grain size, and for the twin thickness $> c$, partial dislocations intersecting with the TB dominate the plastic deformation, which results in strengthening. When $< c$, partial dislocations glide along the TB, which results in softening. Experimentally, Yue et al.[97] directly observed this cross-over in nanocrystalline Cu thin films. As an example, Figures 4e and f present two consecutive HRTEM images collected during the loading. Figures 4g and h are the enlarged HRTEM images collected from the blue-framed region in Figures 4e and f. A partial dislocation resulted in a nucleated and glided incline toward the TB in the thicker twin lamellae (approximately 5 nm). For thinner twin lamella (approximately 1.2 nm), the partial dislocation emission and gliding along the TB resulted in a decrease of the thickness from four atomic layers (Figure 4i) to three and two atomic layers (Figure 4j).

OTHERS: ATOMIC MECHANISMS OF THE DEFORMATION CHARACTERISTICS OF NTS AND HIGH-PRESSURE EXPERIMENTS ON NCS

NTs are a special class of NM because their thickness is typically one or a few atomic layers. The measurements of the Young's modulus of carbon NTs in TEM were firstly conducted by Treacy et al.[98] and by Poncharal et al.[99] Later, many other methods have been used to investigate the plastic deformation mechanisms of carbon NTs. In 2002, Demczyk et al.[100] conducted in situ pulling and bending tests on an individual MWCNT in a TEM. Utilizing the MEMS in the TEM, Zhu et al.[8] measured the fracture strength of MWCNTs, which is 15.84 GPa. Atomic-scale images revealed that the crystalline structure of the MWCNTs (graphite sheets) transformed to amorphous carbon during plastic loading. In 2003, Troiani et al.[101] obtained single-walled carbon nanotubes (SWCNT) from an amorphous C film by the combined effect of irradiation and axial strain. *In situ* HRTEM observations revealed that ductile NTs developed either a junction or a linear chain of C atoms before failure.

Huang et al.[102] discovered that at high temperatures, SWCNT can undergo superplastic deformation. Figure 6a shows a SWCNT with an initial length of 24 nm. At tensile failure, the SWCNT was 91 nm long, which represented a tensile elongation of 280%; its diameter was reduced from 12 to 0.8 nm, as shown in Figure 6d. Kinks frequently form during tensile straining (Figures 6b–d), and they propagate along the tube and then accumulate (Figure 6d) or disappear at the ends. The NT immediately narrows after the kink passes. The super-elongation is due to the high activity of the nucleation and motion of the kink and the atomic diffusion that occurs at high temperatures (2000 °C). The authors' tensile tests at RT revealed that all of the NTs failed at a strain of <15%.

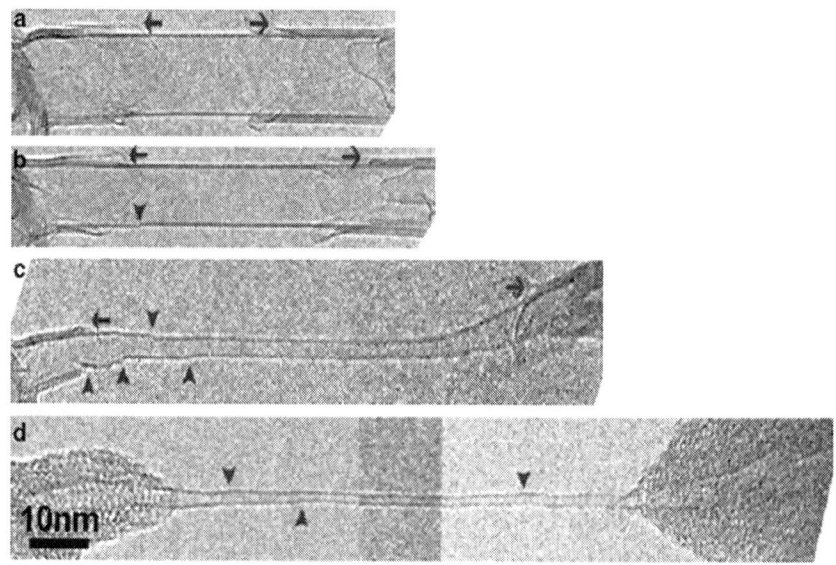

Figure 6: *In situ* tensile elongation of individual single-walled carbon nanotubes viewed via high-resolution transmission electron microscopy. (a–d) Tensile elongation of a single-walled carbon nanotube. Arrowheads mark kinks; arrows indicate features at the ends of the nanotube that are almost unaltered during elongation. Figure from Huang et al.[102] Copyright 2006 Nature Publishing Group.

Because of the special structure, buckling was always observed in the compressive side of the NTs, especially under a bending stress. The atomic-scale structure of this buckling in bent carbon NTs was observed using ex situ HRTEM.[99] Recently, the buckling and fracture modes of thick (diameter >20 nm) MWCNTs under compressive stress were examined using in situ TEM by Zhao et al.[103] The buckling behavior of MWCNTs under compression falls into two categories; the first is non-axial buckling and subsequently complex Yoshimura patterns can be induced on the compressive side of the MWCNTs. The second is axial buckling followed by catastrophic failure. Furthermore, the shell-by-shell fracture mode and planar fracture mode of MWCNTs have been directly observed.

By conducting in situ bending tests in the TEM, Bai et al.[104] observed that multiwalled boron nitride nanotubes (BNNTs) exhibit two interesting phenomena. In situ HRTEM observation revealed that a severely distorted graphitic lattice recovered after the bending strain was released, and the measured I–V curves were suggestive of piezoelectric behavior in the deformed BNNTs, that is, normally electrically insulating multiwalled BNNTs may surprisingly transform to a semiconductor. Using a similar method, Golberg et al.[105] also observed that the multiwalled BNNTs can sustain large strains under bending deformation. The deformation of the BNNTs proceeded through the propagation of consecutive momentary kinks. These kinks were observed to be entirely reversible on reloading with no traces of residual plastic deformation. Low voltage aberration correction TEM makes the almost impossible task of imaging light atoms (boron, carbon, nitrogen or oxygen) without electron beam irradiation damage feasible. However, the deformation experiments of CNT are scarce,[106] and the majority of the works are related to investigations of the static structure or those conducted under electron beam irradiation.

HRTEM is the only method that is capable of directly viewing the atomic structure while providing information about the chemical structure; however, the disadvantage is that the observation must be conducted in a vacuum. It appears almost impossible to directly observe pressure-induced atomic motion. However, Sun et al.[27, 35] developed a new in situ ultra-high pressure experimental method in the TEM. With this method, the atomic-scale behavior of Fe_3C NCs under ultra-high stress was observed in real time.[26] Up to 6% compressive strain was observed in the Fe$_3$C NCs. No visible defects, such as dislocations or twins, were observed during the deformation.

The ultra-high pressure experiment was also conducted using Au NCs,[107] and their TEM results provide evidence that the vacancy concentration in a nanoscale system can be less than in the bulk material. Figures 7a–d shows the gradual extrusion of Au NCs from a graphitic shell at 600 °C. Grain boundaries appeared at the bottleneck in the Au NCs where the deformation rate was the highest (Figures 7e and 7f). When the experimental temperature was ⩽300 °C, twins were occasionally observed. The results indicated that the plastic behaviors are temperature dependent. This conclusion was confirmed by the experiment conducted with Pt at different temperatures[107] and the face-centered cubic phase NCs W and Mo with a typical size of 3–15 nm.[108] The authors claimed that the plastic deformation is governed by the activity of short-lived dislocations and that diffusion of these dislocations may also participate in the plastic deformation. Further experiments and modeling under these special conditions are required.

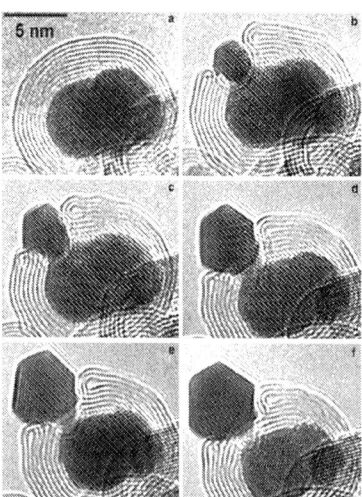

Figure 7: Extrusion of a gold crystal at 600 °C. No visible defects appear in the Au crystal. Irradiation times: (a) 0, (b) 240, (c) 300, (d) 480, (e) 600 and (f) 720 s. Beam current density: 200 A=cm². At 600 °C, the graphitic shells have a high structural perfection. Figure from Sun et al.[107] Copyright 2008 The American Physical Society.

SUMMARY AND PERSPECTIVES

In situ atomic-scale experimental mechanics of NM is a rapidly growing field that is benefitting from technological advancements in instrumentation and improved specimen fabrication techniques, although stress–strain correlations are still absent in the majority of current studies. For nano-mechanics, *in situ* TEM experiments provide the possibility to directly observe the deformation mechanisms along with measuring stress–strain data at the appropriate length scale and dynamic observations, which are essential for understanding the deformation mechanisms of NMs. In the past decades, significant progress has been made regarding the deformation dynamics of NMs using quantified techniques, and the influence of material dimensions on the nucleation and multiplication of dislocations has, to some extent, been resolved. However, open questions remain regarding the dynamics of dislocation interactions in small volumes. With the emergence of atomic-scale experimental nano-mechanics, new opportunities and research fields may arise in this regime.

Furthermore, with the integration of equipment that allows for the quantification of strain–stress output for small-sized NM with HRTEM, Cs-corrected atomic-scale HRTEM imaging, individual atomic probing ability and real-time and atomic-scale chemical information detection, a golden age for conducting investigations of the experimental mechanics of materials at atomic scale can be expected, and the related research field will be very optimistic in the near future. However, care should be taken when conducting these *in situ* deformation dynamics studies using TEM due to the shower of electron beam irradiation. By taking advantage of the electron beam irradiation, unusual mechanical properties can be approached.[37,109,110] However, with an overdose of this radiation, significant irradiation effects will be integrated with the intrinsic physical properties of these materials. When conducting *in situ* experimental microscopy mechanics or physics investigations by assigning the fast electrons to interact with the observed objects, we have to obtain a balance between 'watching' and 'damaging' (overdosing).

Regarding atomic-scale nano-mechanics, to be specific, it is optimistic that several research fields can find solutions from the integrated technology of stress–strain correlations, atom-by-atom structural, chemical and even electronic property investigation with external stress/straining on the materials:

- For complex nanosystems that consist of multiple elements, atomic-scale imaging with accurate chemical distinction ability is necessary. Under stress/straining, the time-resolved and dynamic atomic-scale imaging ability for the strained materials provide new opportunities in nanoscience;
- The direct atomic mechanisms of GB sliding, diffusion and rotation for complex alloys (multiple elements and/or intermetallic compounds) are highly important for developing novel nanocrystalline materials;
- The interface structural evolution process includes the direct atomic-scale understanding of the twinning nucleation and/or propagation process in nanocrystalline materials;
- The interphase interface structural evolution process of those multiple-phased NMs when they are under stress/strain;
- With the ability of ultra-large elasticity, the lattice spacing of the NM can be significantly changed, and the band structure and the related physical and chemical properties can be tuned accordingly. The new fields of strained/stressed engineering are promising for NMs because the strain-induced polarity of NM creates novel piezoelectronics, giant magnetic-resistance effects, phase transitions, and so on. All of these needs not only require atomic-scale understanding but also Cs-corrected imaging to directly map the polarity, atomic-scale strain distributions with single atom resolution and accuracy (even for light atoms such as C, O and N atoms, and so on). New materials, novel functionalities and applications can thus be designed with atomic-scale precision. With, but not limited to the aforementioned possibilities, a golden age of 'nano-mechanics with an atomic scale' is promising.

ACKNOWLEDGEMENTS

This work was supported by the National 973 Program of China (2009CB623700), the Key Project of C-NSF (11127404, 50831001), the Beijing High-level Talents (PHR20100503), the Beijing

PXM201101420409000053, the Beijing 211 Project and the National Outstanding Young Scientist Grant of China (10825419).

REFERENCES

1. Wong, E. W., Sheehan, P. E. & Lieber, C. M. Nanobeam mechanics: elasticity, strength, and toughness of nanorods and nanotubes. *Science* 277, 1971–1974 (1997).
2. Wu, B., Heidelberg, A. & Boland, J. J. Mechanical properties of ultrahigh-strength gold nanowires. *Nat. Mater.* 4, 525–529 (2005).
3. Yu, M., Lourie, O., Dyer, M. J., Moloni, K., Kelly, T. F. & Ruoff, R. S. Strength and breaking mechanism of multiwalled carbon nanotubes under tensile load. *Science* 287, 637–640 (2000).
4. Shan, Z. W., Adesso, G., Cabot, A., Sherburne, M. P., Asif, S. A. S., Warren, O. L., Chrzan, D. C., Minor, A. M. & Alivisatos, A. P. Ultrahigh stress and strain in hierarchically structured hollow nanoparticles. *Nat. Mater.* 7, 947–952 (2008).
5. Uchic, M. D., Dimiduk, D. M., Florando, J. N. & Nix, W. D. Sample dimensions influence strength and crystal plasticity. *Science* 305, 986–989 (2004).
6. Shan, Z. W., Mishra, R. K., Asif, S. A. S., Warren, O. L. & Minor, A. M. Mechanical annealing and source-limited deformation in submicrometre-diameter Ni crystals. *Nat. Mater.* 7, 115–119 (2008).
7. Kiener, D. & Minor, A. M. Source truncation and exhaustion: insights from quantitative *in situ* TEM tensile testing. *Nano Lett.* 11, 3816–3820 (2011).
8. Zhu, Y. & Espinosa, H. D. An electromechanical material testing system for *in situ* electron microscopy and applications. *Proc. Natl Acad. Sci. USA* 102, 14503–14508 (2005).
9. Kumar, K. S., Swygenhoven, H. V. & Suresh, S. Mechanical behavior of nanocrystalline metals and alloys. *Acta Mater.* 51, 5743–5774 (2003).
10. Zhu, T. & Li, J. Ultra-strength materials. *Prog. Mater. Sci.* 55, 710–757 (2010).

11. Swygenhoven, H. V. Grain boundaries and dislocations. *Science* 296, 66–67 (2002).
12. Schiotz, J., Di Tolla, F. D. & Jacobsen, K. W. Softening of nanocrystalline metals at very small grain sizes. *Nature* 391, 561–563 (1998).
13. Espinosa, H. D., Bernal, R. A. & Filleter, T. In situ TEM electromechanical testing of nanowires and nanotubes. *Small* 8, 3233–3252 (2012).
14. Agrawal, R., Loh, O. & Espinosa, H. D. The evolving role of experimental mechanics in 1-D nanostructure-based device development. *Exp. Mech.* 51, 1–9 (2011).
15. Motz, C. in *Handbook of In-Situ Electron Microscopy: Applications in Physics, Chemistry and Materials Science* (eds Dehm, G., Howe, J. M. & Zweck, J.) Ch. 9, 193–252 (Wiley, Germany, 2012).
16. Jolandan, M. M., Bernal, R.A., Kuljanishvili, I., Parpoil, V. & Espinosa, H.D. Individual GaN nanowires exhibit strong piezoelectricity in 3D. *Nano Lett.* 12, 970–976 (2012).
17. Hirsch, P., Howie, A., Nicholson, R., Pashley, D. W. & Whelan, M. J. *Electron microscopy of thin crystals* (Krieger Publishing Company, Malabar, 1967).
18. Zhu, Y., Qin, Q. Q., Xu, F., Fan, F. G., Ding, Y., Zhang, T., Wiley, B. J. J. & Wang, Z. L. Size effects on elasticity, yielding, and fracture of silver nanowires: *in situ* experiments. *Phys. Rev. B.* 85, 045443 (2012).
19. Li, N., Wang, J., Huang, J. Y., Misra, A. & Zhang, X. Influence of slip transmission on the migration of incoherent twin boundaries in epitaxial nanotwinned Cu. *Scripta Mater.* 64, 149–152 (2011).
20. Ke, M., Hackney, S. A., Milligan, W. W. & Aifantis, E. C. Observation and measurement of grain rotation and plastic strain in nanostructured metal thin films. *Nanotruct. Mater.* 5, 689–697 (1995).
21. Oh, S. H., Legros, M., Kiener, D. & Dehm, G. In situ observation of dislocation nucleation and escape in a submicrometre aluminium single crystal. *Nat. Mater.* 8, 95–100 (2009).
22. Agrait, N., Rodrigo, J. G. & Vieira, S. Conductance steps and quantization in atomic-size contacts. *Phys. Rev. B.* 47, 12345–12348 (1993).

23. Lin, L. T., Cui, T. R., Qin, L. C. & Washburn, S. Direct measurement of the friction between and shear moduli of shells of carbon nanotubes. *Phys. Rev. Lett.* 107, 206101 (2011).
24. Tang, D. M., Yin, L. C., Li, F., Liu, C., Yu, W. J., Hou, P. X., Wu, B., Lee, Y. H., Ma, X. L. & Cheng, H. M. Carbon nanotube-clamped metal atomic chain. *Proc. Natl Acad. Sci. USA* 107, 9055–9059 (2010).
25. Han, X. D., Zheng, K., Zhang, Y. F., Zhang, X. N., Zhang, Z. & Wang, Z. L. Low-temperature *in situ* large-strain plasticity of silicon nanowires. *Adv. Mater.* 19, 2112–2118 (2007).
26. Han, X. D., Zhang, Y. F., Zheng, K., Zhang, X. N., Zhang., Z., Hao, Y. J., Guo, X. Y., Yuan, J. & Wang, Z. L. Low-temperature *in situ* large strain plasticity of ceramic SiC nanowires and its atomic-scale mechanism. *Nano Lett.* 7, 452–457 (2007).
27. Sun, L. T., Banhart, F., Krasheninnikov, A. V., Rodríguez-Manzo, J. A., Terrones, M. & Ajayan, P. M. Carbon nanotubes as high-pressure cylinders and nanoextruders. *Science* 312, 1199–1202 (2004).
28. Zhang, Y. F., Han, X. D., Zheng, K., Zhang, Z., Zhang, X. N., Fu, J. Y., Ji, Y., Hao, Y. J., Guo, X. Y. & Wang, Z. L. Direct observation of super-plasticity of beta-SiC nanowires at low temperature. *Adv. Func. Mater.* 17, 3435–3440 (2007).
29. Wang, L. H., Han, X. D., Liu, P., Yue, Y. H., Zhang, Z. & Ma, E. *In situ* observation of dislocation behavior in nanometer grains. *Phys. Rev. Lett.* 105, 135501 (2010).
30. Haque, M. A. & Saif, M. T. A. In-situ tensile testing of nano-scale specimens in SEM and TEM. *Exp. Mech.* 42, 123–128 (2002).
31. Legros, M., Gianola, D. S. & Hemker, K. J. In situ TEM observations of fast grain-boundary motion in stressed nanocrystalline aluminum films. *Acta Mater.* 56, 3380–3393 (2008).
32. Mompioua, F., Legrosa, M., Sedlmayrb, A., Gianolac, D. S., Caillarda, D. & Kraft, O. Source-based strengthening of sub-micrometer Al fibers. *Acta Mater.* 60, 977–983 (2006).
33. Kiener, D., Hosemann, P., Maloy, S. A. & Minor, A. M. In situ nanocompression testing of irradiated copper. *Nat. Mater.* 10, 608–613 (2011).

34. Minor, A. M., Asif, S. A., Shan, Z., Stach, E. A., Cyrankowski, E., Wyrobek, T. J. & Warren, O. L. A new view of the onset of plasticity during the nanoindentation of aluminium. *Nat. Mater.* 5, 697–702 (2006).
35. Sun, L. T., Rodríguez-Manzo, J. A. & Banhart, F. Elastic deformation of nanometer-sized metal crystals in graphitic shells. *Appl. Phys. Lett.* 89, 263104 (2006).
36. Zheng, K., Han, X. D., Wang, L. H., Zhang, Y. F., Yue, Y. H., Qin, Y., Zhang, X. N. & Zhang, Z. Atomic mechanisms governing the elastic limit and the incipient plasticity of bending Si nanowires. *Nano Lett.* 9, 2471–2476 (2009).
37. Zheng, K., Wang, C. C., Cheng, Y. Q., Yue, Y. H., Han, X. D., Zhang, Z., Shan, Z. W., Mao, S. X., Ye, M. M., Yin., Y. D. & Ma, E. Electron-beam-assisted superplastic shaping of nanoscale amorphous silica. *Nat. Comm.* 1, 24 (2010).
38. Han, X. D., Wang, L. H., Liu, P., Yue, Y. H., Yang, M. J., Sun, J. L. & Zhang, Z. Dynamic atomic mechanisms of plasticity of Ni nanowires and nano crystalline ultra-thin films. *Mater. Scien. Forum* 654, 2293–2296 (2010).
39. Dehm, G. Miniaturized single-crystalline fcc metals deformed in tension: new insights in size-dependent plasticity. *Prog. Mater. Sci.* 54, 664–688 (2009).
40. Legros, M., Dehm, G., Keller-Flaig, R. M., Arzt, E., Hemker, K. J. & Suresh, S. Dynamic observation of Al thin films plastically strained in a TEM. *Mater. Sci. Eng. A* 15, 463–467 (2001).
41. Wang, L. H., Zhang, Z., Ma, E. & Han, X. D. Transmission electron microscopy observations of dislocation annihilation and storage in nanograins. *Appl. Phys. Lett.* 98, 051905 (2011).
42. Liu, P., Mao, S. C., Wang, L. H., Han, X. D. & Zhang, Z. Direct dynamic atomic mechanisms of strain-induced grain rotation in nanocrystalline, textured, columnar-structured thin gold films. *Scripta Mater.* 64, 343–346 (2011).
43. Yue, Y. H., Liu, P., Zhang, Z., Han, X. D. & Ma, E. Approaching the theoretical elastic limit in copper nanowires. *Nano Lett.* 11, 3151–3155 (2011).
44. Yue, Y. H., Liu, P., Deng, Q. S., Ma, E., Zhang, Z. & Han, X. D. Quantitative evidence of crossover toward partial dislocation

mediated plasticity in copper single crystalline nanowires. *Nano Lett.* 12, 4045–4049 (2012).

45. Han, X. D., Zhang, Z. & Wang, Z. L. Experimental nanomechanics of one-dimensional nanomaterials by *in situ* microscopy. *Nano* 2, 249–271 (2007).
46. Deng, Q. S., Cheng, Y. Q., Yue, Y. H., Zhang, L., Zhang, Z., Han, X. D. & Ma, E. Uniform tensile elongation in framed submicron metallic glass specimen in the limit of suppressed shear banding. *Acta Mater.* 59, 6511–6518 (2011).
47. Jiang, Q. K., Liu, P., Ma, Y., Cao, Q. P., Wang, X. D., Zhang, D. X., Han, X. D., Zhang, Z. & Jiang, J. Z. Super elastic strain limit in metallic glass films. *Sci. Rep.* 2, 852 (2012).
48. Taylor, G. F. A method of drawing metallic filaments and a discussion of their properties and uses. *Phys. Rev.* 23, 655–660 (1924).
49. Bragg, L. & Lomer, W. M. A dynamical model of a crystal structure ii. *Proc. R. Soc. Lond. A* 196, 171–181 (1949).
50. Herring, C. & Galt, J. K. Elastic and plastic properties of very small metal specimens. *Phys. Rev.* 85, 1060–1061 (1952).
51. Deneen, J., Mook, W. M., Minor, A., Gerberich, W. W. & Carter, C. B. In situ deformation of silicon nanospheres. *J. Mater. Sci.* 41, 4477–4483 (2006).
52. Uchic, M. D., Shade, P. A. & Dimiduk, D. M. Plasticity of micrometer-scale single crystals in compression. *Annu. Rev. Mater. Res.* 39, 361–386 (2009).
53. Kraft, O., Gruber, P. A., Monig, R. & Weygand, D. Plasticity in confined dimensions. *Annu. Rev. Mater. Res.* 40, 293–317 (2010).
54. Dou, R. & Derby, B. A universal scaling law for the strength of metal micropillars and nanowires. *Scripta Mater.* 61, 524–527 (2009).
55. Greer, J. R. & De Hosson, J. M. Plasticity in small-sized metallic systems: Intrinsic versus extrinsic size effect. *Prog. Mater. Sci.* 56, 654–724 (2011).
56. Greer, J. R., Oliver, W. C. & Nix, W. D. Size dependence of mechanical properties of gold at the micron scale in the absence of strain gradients. *Acta Mater.* 53, 1821–1830 (2005).

57. Kiener, D., Grosinger, W., Dehm, G. & Pippan, R. A further step towards an understanding of size-dependent crystal plasticity: *in situ* tension experiments of miniaturized single-crystal copper samples. *Acta Mater.* 56, 580–592 (2008).
58. Dehm, G., Motz, C., Scheu, C., Clemens, H., Mayrhofer, P. H. & Mitterer., C. Mechanical size-effects in miniaturized and bulk materials. *Adv. Eng. Mater.* 8, 1033–1045 (2006).
59. Legros, M., Gianola, D. S. & Motz, C. Quantitative *in situ* mechanical testing in electron microscopes. *MRS Bull.* 35, 354–360 (2010).
60. Greer, J. R. & Nix, W. D. Nanoscale gold pillars strengthened through dislocation starvation. *Phys. Rev. B.* 73, 245410 (2006).
61. Parthasarathy, T. A., Rao, S. I., Dimiduk, D. M., Uchic, M. D. & Trinkle, D. R. Contribution to size effect of yield strength from the stochastics of dislocation source lengths in finite samples. *Scripta Mater.* 56, 313–316 (2007).
62. Seo, J. H., Yoo, Y. D., Park, N. Y., Yoon, S. W., Lee, H., Han, S., Lee, S. W., Seong, T. Y., Lee, S. C., Lee, K. B., Cha, P. R., Park, H. S., Kim, B. & Ahn, J. P. Superplastic deformation of defect-free Au nanowires via coherent twin propagation. *Nano Lett.* 11, 3499–3502 (2011).
63. Sedlmayr, A., Bitzek, E., Gianola, D., Richter, S., Mönig, G. & Kraft, R. O. Existence of two twinning-mediated plastic deformation modes in Au nanowhiskers. *Acta Mater.* 60, 3985–3993 (2012).
64. Lu, Y., Song, J., Huang, J. Y. & Lou, J. Fracture of sub-20 nm ultrathin gold nanowires. *Adv. Funct. Mater.* 21, 3982–3989 (2011).
65. Oh, S. H., Legros, M., Kiener, D., Gruber, P. & Dehm, G. In situ TEM straining of single crystal Au films on polyimide: Change of deformation mechanisms at the nanoscale. *Acta Mater.* 55, 5558–5571 (2007).
66. Zheng, H., Cao, A. J., Weinberger, C. R., Huang, J. Y., Du, K., Wang, J. B., Ma, Y. Y., Xia, Y. N. & Mao, S. X. Discrete plasticity in sub-10 nm-sized gold crystals. *Nat. Commun.* 1, 144 (2010).
67. Kizuka, T. Atomistic visualization of deformation in gold. *Phys. Rev. B.* 57, 11158–11163 (1998).

68. Kizuka, T. Atomic process of point contact in gold studied by time-resolved high-resolution transmission electron microscopy. *Phys. Rev. Lett.* 81, 4448–4451 (1998).
69. Matsuda, T. & Kizuka, T. Structure of nanometer-sized palladium contacts and their mechanical and electrical properties. *Jpn J. Appl. Phys.* 46, 4370–4374 (2007).
70. Nie, A. & Wang, H. T. Deformation-mediated phase transformation in gold nano-junction. *Mater. Lett.* 65, 3380–3383 (2011).
71. Ohnishi, H., Kondo, Y. & Takayanagi, K. Quantized conductance through individual rows of suspended gold atoms. *Nature* 395, 780–783 (1998).
72. Bettini, J., Sato, F., Coura, P. Z., Dantas, S. O., Galva, D. S. & Ugarte, D. Experimental realization of suspended atomic chains composed of different atomic species. *Nat. Nanotechnol.* 1, 182–185 (2006).
73. Michler, J., Wasmer, K., Meier, S., Östlund, F. & Leifer, K. Plastic deformation of gallium arsenide micropillars under uniaxial compression at room temperature. *Appl. Phys. Lett.* 90, 043123 (2007).
74. Landau, L. D. & Lifshitz, E. M. *Theory of Elasticity* 64-86 (Pergamon Press, New York, 1986).
75. Smith, D. A., Holmberg, V. C. & Korgel, B. A. Flexible germanium nanowires: ideal strength, room temperature plasticity, and bendable semiconductor fabric. *ACS Nano* 4, 2356–2362 (2010).
76. Asthana, A., Momeni, K., Prasad, A., Yap, Y. K. & Yassar, R. S. In situ observation of size-scale effects on the mechanical properties of ZnO nanowires. *Nanotechnology* 22, 265712 (2011).
77. Tang, D. M., Ren, C. L., Wang, M. S., Wei, X. L., Kawamoto, N., Liu, C., Bando, Y., Mitome, M., Fukata, N. & Golberg, D. Mechanical properties of Si nanowires as revealed by *in situ* transmission electron microscopy and molecular dynamics simulations. *Nano Lett.* 12, 1898–1904 (2012).
78. Östlund, F., Malyska, K. R., Leifer, K., Hale, L. M., Tang, Y. Y., Ballarini, R., Gerberich, W. W. & Michler, J. Brittle-to-ductile transition in uniaxial compression of silicon pillars at room temperature. *Adv. Funct. Mater.* 19, 2439–2444 (2009).

79. Ostlund, F., Howie, P. R., Ghisleni, R., Korte, S., Leifer, K., Clegg, W. J. & Michler, J. Ductile-brittle transition in micropillar compression of GaAs at room temperature. *Phil. Mag.* 91, 1190–1199 (2011).
80. Stauffer, D. D., Beaber, A., Wagner, A., Ugurlu, O., Nowak, J. L., Mkhoyan, K. A., Girshick, S. & Gerberich, W. W. Strain-hardening in submicron silicon pillars and spheres. *Acta Mater.* 60, 2471–2478 (2012).
81. Nowak, J. D., Beaber, A. R., Ugurlu, O., Girshickd, S. L. & Gerberich, W. W. Small size strength dependence on dislocation nucleation. *Scripta Mater.* 62, 819–822 (2010).
82. Mook, W. M., Nowak, J. D., Perrey, C. R., Carter, C. B., Mukherjee, R., Girshick, S. L., McMurry, H. & Gerberich, W. W. Compressive stress effects on nanoparticle modulus and fracture. *Phys. Rev. B.* 75, 214112 (2007).
83. Wang, L. H., Zheng, K., Zhang, Z. & Han, X. D. Direct atomic-scale imaging about the mechanisms of ultralarge bent straining in Si nanowires. *Nano Lett.* 11, 2382–2385 (2011).
84. Hirth, J. P. & Lothe, J. *Theory of Dislocations* 217-834 (Wiley, New York, 1992).
85. Meyers, M. A., Mishra, A. & Benson, D. J. Mechanical properties of nanocrystalline materials. *Prog. Mater. Sci.* 51, 427–556 (2006).
86. Shan, Z. W., Stach, E. A., Wiezorek, J. M. K., Knapp, J. A., Follstaedt, D. M. & Mao, S. X. Grain boundary–mediated plasticity in nanocrystalline nickel. *Science* 305, 654–657 (2004).
87. Jin, M., Minor., A. M., Stach, E. A. & Morris, J. W. Direct observation of deformation-induced grain growth during the nanoindentation of ultrafine-grained Al at room temperature. *Acta Mater.* 52, 5381 (2004).
88. Youssef, K. M., Scattergood, R. O., Murty, K. L., Horton, J. A. & Koch, C. C. Ultrahigh strength and high ductility of bulk nanocrystalline copper. *Appl. Phys. Lett.* 87, 091904 (2005).
89. Li, B. Q., Sui, M. L. & Mao, S. X. Pseudoelastic stacking fault and deformation twinning in nanocrystalline Ni. *Appl. Phys. Lett.* 97, 241912 (2010).
90. Chen, M. W., Ma, E., Hemker, K. J., Sheng, H. W., Wang, Y. M. & Cheng, X. M. Deformation twinning in nanocrystalline aluminum. *Science* 300, 1275–1277 (2003).

91. Li, B. Q., Li, B., Wang, Y. B., Sui, M. L. & Ma., E. Twinning mechanism via synchronized action of partial dialocations in face-centered-cubic materials.*Scripta Mater.* 64, 852–855 (2011).
92. Wu, X. L., Liao, X. Z., Srinivasan, S. G., Zhou, F., Lavernia, E. J., Valiev, R. Z. & Zhu, Y. T. New deformation twinning mechanism generates zero macroscopic strain in nanocrystalline metals. *Phys. Rev. Lett.* 100, 095701 (2008).
93. Lu, L., Chen, X., Huang, X. & Lu, K. Revealing the maximum strength in nanotwinned copper. *Science* 323, 607–610 (2009).
94. Wang, Y. B. & Sui, M. L. Atomic-scale *in situ* observation of lattice dislocations passing through twin boundary. *Appl. Phys. Lett.* 94, 021909 (2009).
95. Wang, J., Li, N., Anderoglu, O., Zhang, X., Misra, A., Huang, J. Y. & Hirth, J. P. Detwinning mechanisms for growth twins in face-centered cubic metals.*Acta Mater.* 58, 2262–2270 (2010).
96. Li, X. Y., Wei, Y. J., Lu, L., Lu, K. & Gao, H. J. Dislocation nucleation governed softening and maximum strength in nano-twinned metals. *Nature* 464, 877–880 (2010).
97. Yue, Y. H., Wang, L. H., Zhang, Z. & Han, X. D. Cross-over of the plasticity mechanism in nanocrystalline Cu. *Chin. Phys. Lett.* 29, 066201 (2012).
98. Treacy, M. M. J., Ebbesen, T. W. & Gibson, J. M. Exceptionally high Young's modulus observed for individual carbon nanotubes. *Nature* 381, 678–680 (1996).
99. Poncharal, P., Wang, Z. L., Ugarte, D. & de Heer, W. A. Electrostatic deflections and electromechanical resonances of carbon nanotubes. *Science*283, 1513–1516 (1999).
100. Demczyk, B. G., Wang, Y. M., Cumings, J., Hetman, M., Han, W., Zettl, A. & Ritchie, R. O. Direct mechanical measurement of the tensile strength and elastic modulus of multiwalled carbon nanotubes. *Mater. Sci. Eng. A* 334, 173–178 (2002).
101. Troiani, H. E., Yoshida, M. M., Camacho, B. G. A., Marques, M. A. L., Rubio, A., Ascencio, J. A. & Yacaman, M. J. Direct observation of the mechanical properties of single-walled carbon nanotubes and their junctions at the atomic level. *Nano Lett.* 3, 751–755 (2003).

102. Huang, J. Y., Chen, S., Wang, Z. Q., Kempa, K., Wang, Y. M., Jo, S. H., Chen, G., Dresselhaus, M. S. & Ren, Z. F. Superplastic carbon nanotubes. *Nature* 439, 281 (2006).
103. Zhao, J., He, M. R., Dai, S., Huang, J. Q., Wei, F. & Zhu, J. TEM observations of buckling and fracture modes for compressed thick multiwall carbon nanotubes. *Carbon* 49, 206–213 (2011).
104. Bai, X. D., Golberg, D., Bando, Y., Zhi, C., Tang, C. C., Mitome, M. & Kurashima, K. Deformation-driven electrical transport of individual boron nitride nanotubes. *Nano Lett.* 7, 632–637 (2007).
105. Golberg, D., Costa, P. F. J., Lourie, O., Mitome, M., Bai, X. D., Kurashima, K. J., Zhi, C. Y., Tang, C. C. & Bando, Y. S. Direct force measurements and kinking under elastic deformation of individual multiwalled boron nitride nanotubes. *Nano Lett.* 7, 2146–2151 (2007).
106. Warner, J. H., Young, N. P., Kirkland, A. I. & Briggs, G. A. D. Resolving strain in carbon nanotubes at the atomic level. *Nat. Mater.* 10, 958–962 (2011).
107. Sun, L. T., Krasheninnikov, A. V., Ahlgren, T., Nordlund, K. & Banhart, F. Plastic deformation of single nanometer-sized crystals. *Phys. Rev. Lett.* 101, 156101 (2008).
108. Li, J. X. & Banhart, F. The deformation of single, nanometer-sized metal crystals in graphitic shells. *Adv. Mater.* 17, 1539–1542 (2005).
109. Wang, L. H., Han, X. D., Zhang, Y. F., Zheng, K., Liu, P. & Zhang, Z. Asymmetrical quantum dot growth on tensile and compressive-strained ZnO nanowire surfaces. *Acta Mater.* 59, 651–657 (2011).
110. Dai, S., Zhao, J., Xie, L., Cai, Y., Wang, N. & Zhu, J. Electron-beam-induced elastic–plastic transition in Si nanowires. *Nano Lett.* 12, 2379 (2012).

# Chapter 7

Rate Sensitive Continuum Damage Models and Mesh Dependence in Finite Element Analyses

Goran Ljustina, Martin Fagerström, and Ragnar Larsson

Division of Material and Computational Mechanics, Department of Applied Mechanics, Chalmers University of Technology, 412 96 Gothenburg, Sweden

ABSTRACT

The experiences from orthogonal machining simulations show that the Johnson-Cook (JC) dynamic failure model exhibits significant element size dependence. Such mesh dependence is a direct consequence of the utilization of local damage models. The current contribution is an investigation of the extent of the possible pathological mesh dependence. A comparison of the resulting JC model behavior combined with two types of damage evolution is considered. The first damage model is the JC dynamic failure model, where the development of the "damage" does not affect the response until the critical state is reached. The second one is a continuum damage model, where the damage variable is affecting the material response continuously during the deformation. Both the plasticity and the damage models are rate dependent, and the damage evolutions for both models are defined as a postprocessing of the effective stress response. The investigation is conducted for a series of 2D shear tests utilizing different FE representations of the plane strain plate with pearlite material properties. The results show for both damage models, using realistic pearlite material parameters, that similar extent of the mesh dependence is obtained and that the possible viscous regularization effects are absent in the current investigation.

INTRODUCTION

The traditional Johnson-Cook (JC) plasticity model [1] is often used in finite element (FE) analysis of metal cutting. From the industrial perspective, one important advantage of the JC model, compared to many others, is its availability as built in constitutive model in the commercial software packages. In the JC model, it is assumed that the flow stress is a unique function of the total strain, plastic strain rate, and temperature and that their effects on the flow stress can be described in a multiplicative fashion.

The accuracy of phenomenological models, like the JC plasticity model, is often satisfactory in the range of deformation conditions for which they were curve fitted. What is missing is the ability to capture the kinematic hardening, recovery, or complex loading mechanisms that are common in machining. On the other hand, as one example of dislocation mechanics based models, the BCJ (Bammann-Chiesa-Johnson) model [2, 3] has been developed to incorporate the loading effect in complex material deformations and is successfully used earlier in orthogonal machining simulations of the cast iron material (cf. [4, 5]).

Machining simulation models in many cases include damage models in order to describe the development of segmented or discontinuous chips (cf. [6–11]). In the earlier contributions [12, 13], a Johnson-Cook (JC) plasticity model was used along with the Johnson-Cook dynamic failure criterion to describe deformation and damage in 2D orthogonal machining simulations with a FE-resolved compacted graphite iron (CGI) microstructure.

The JC dynamic failure [14, 15] defines a simple damage/failure model when the "damage" development in simulations is manifested through an accompanying element deletion procedure. The element is deleted when the accumulated plastic strain reaches a critical value, and there is thus no "damage" influence on the stress response before the critical plastic strain is reached and the element is deleted. This critical or failure strain is dependent on a nondimensional plastic strain rate, stress triaxiality, and temperature. Hence, due to the fact that the Johnson-Cook dynamic failure model is not affecting the stress response until the failure strain is reached, the model will be referred to as the "uncoupled damage" (UD) model from here on. Despite the fact that rate dependence is included in the Johnson-Cook plasticity model, and thereby a certain "viscous regularization" [16, 17] can be expected, our experience with machining simulations using this model indicates the presence of the pathological mesh dependence [18, 19]. Thereby, the aim of the present paper is to investigate the significance of the pathological mesh dependence obtained for different FE representations of the plane strain plate with pearlite material properties.

Another failure model that has been successfully used in machining simulations of cast iron with resolved microstructure in the literature (cf. [4]) is the damage model describing spherical voids growth by Cocks and Ashby [20]. In contrast to the JC dynamic failure model, the effective stress response is in this model reduced by a scalar damage variable in the plastic deformation phase creating a more realistic stress-strain pattern. Unlike the uncoupled damage model that requires five material parameters (d_1–d_5), one single material parameter is needed for this model. This damage model is referred to as the "continuum damage" (CD) model in this paper. So, the additional objective of this contribution is to compare the two damage models with respect to pathological mesh dependence.

DAMAGE MODELS BASED ON VISCOPLASTIC EVOLUTION

In this section we describe two fundamentally different damage/failure models, used to represent ductile fracture of the pearlite material. In the continuum damage model, the evolving damage is affecting the response progressively as the plastic deformation develops, whereas in the uncoupled damage model the evolved damage takes place in one single step when the plastic strain reaches a critical value. Both models rely on the Johnson-Cook model for the representation of the effective stress response.

Effective Material and the Johnson-cook Model

It is assumed that the Helmholtz free energy $\bar{\psi}$ of the effective (undamaged) material, denoted by a superimposed bar, has the dependencies

$$\bar{\psi} = \bar{\psi}[\mathbf{C}, k], \quad (1)$$

where $C=F^t.F$ (where F is the deformation gradient) is the elastic right Cauchy-Green deformation tensor and is the isotropic hardening variable. It then appears that the dissipation rate $\bar{D}$ of the dissipation inequality becomes

$$\bar{\mathcal{D}} = \bar{\tau} : \mathbf{1}_p + \bar{\kappa}\dot{k} \geq 0 \quad (2)$$

corresponding to the constitutive state equationsIn

$$\bar{S} = \frac{\partial \bar{\psi}}{\partial \mathbf{C}} \Longrightarrow \bar{\tau} = \mathbf{F} \cdot \bar{\mathbf{S}} \cdot \mathbf{F}^t, \qquad \bar{\kappa} = -\frac{\partial \bar{\psi}}{\partial k}. \quad (3)$$

In (3), we introduced the effective Kirchhoff stress $\bar{\tau}$ related to the effective second Piola Kirchhoff stress $\bar{S}$ with the usual transformation. Moreover, $\bar{k}$ is the hardening stress pertinent to the effective material. In view of (2), let us introduce the evolution of the internal dissipative variables $\{l_p, k\}$ in terms of the yield function of the effective stress; that is, $\phi = \phi[\bar{\tau}, \bar{k}]$ The evolution of the internal variables is thus formulated as

$$\mathbf{l}_p = \lambda \frac{\partial \Phi}{\partial \bar{\tau}} = \lambda \mathbf{f} \Longrightarrow \mathbf{f} = \frac{\partial \Phi}{\partial \bar{\tau}} = \frac{3}{2} \frac{\bar{\tau}_{dev}}{\bar{\tau}_e}, \quad \dot{k} = \lambda \frac{\partial \Phi}{\partial \kappa},$$

(4)

where $\lambda \geq 0$ is the plastic multiplier (pertinent to the effective inelastic response) and f is the gradient of the yield function in terms of the effective Kirchhoff stress $\bar{\tau}$. As alluded to in the yield function $\phi = \phi[\bar{\tau}]$, von Mises plasticity is considered involving the von Mises stress $\bar{\tau}_e$ defined by

$$\bar{\tau}_e = \sqrt{\frac{3}{2}} |\bar{\tau}_{dev}|.$$

(5)

As to the plastic multiplier $\lambda \geq 0$, let us consider the Johnson-Cook model [1] where the overstress function is specified in the quasi-static yield function as

$$\lambda = \begin{cases} \dot{\epsilon}_0 \exp\left[\dfrac{\Phi}{C\left(1-\hat{\theta}^m\right)(A+Bk^n)}\right] & \dfrac{\lambda}{\dot{\epsilon}_0} \geq 1 \\ \Phi \leq 0, \ \lambda \geq 0, \ \lambda\Phi = 0 & \dfrac{\lambda}{\dot{\epsilon}_0} < 1, \end{cases}$$

(6)

where $k = \epsilon_e^p$ is the internal hardening variable (equal to the equivalent effective plastic strain) and $\dot{\lambda} = \dot{k} = \dot{\epsilon}_e^p$. Furthermore, the quasi-static yield function is defined as

$$\Phi = \bar{\tau}_e - \left(A + B(k)^n\right)\left(1 - \hat{\theta}^m\right).$$

(7)

We conclude that the response becomes rate dependent (or viscoplastic) whenever $\lambda \geq \dot{\epsilon}_0$. The response becomes rate independent whenever $\lambda < \dot{\epsilon}_0$ controlled by usual loading conditions introduced in (6). Moreover, the factor $(1-\hat{\theta}^m)$ involves the temperature dependence where $\hat{\theta}$ is the homologous temperature and m is a parameter (cf. [1, 13]). As to the material parameters involved, we note that A, B and C are material parameters representing initial yield, hardening, and rate sensitivity, respectively. In addition, the exponent n represents the hardening. Concerning the parameter $\dot{\varepsilon}_o$ we note that it has a strong influence on the rate sensitivity.

A Damage Enhanced Formulation

In the present context it is of significant interest to enhance the effective material in terms of a scalar (isotropic) damage measure ϕ acting on the effective material so that

$$\psi = (1 - \phi)\overline{\psi} \qquad (8)$$

whereby the total dissipation rate D becomes extended as

$$\mathcal{D} = (1 - \phi)\overline{\mathcal{D}} + \overline{\psi}\dot{\phi} \geq 0, \qquad (9)$$

where (again) $\overline{D}$ is the effective dissipation rate. To ensure positive dissipation $D \geq 0$ it suffices to consider $\overline{D} \geq 0$ as in the sequel (2)–(4) and $\dot{\phi} \geq 0$ (since $\overline{\psi} \geq 0$ always). This corresponds to the constitutive state equations:

$$\tau = (1 - \phi)\overline{\tau}, \qquad \kappa = (1 - \phi)\overline{\kappa},$$

$$A = -\frac{\partial \psi}{\partial \phi} = \overline{\psi}. \qquad (10)$$

We thus conclude that, for example, the total Kirchhoff stress is obtained via the relation $\tau = (1 - \phi)\overline{\tau}$ and that A is the damage force driving the damage evolution via the elastically stored effective energy

A Continuum Damage Model

The structure proposed in the previous subsection involves a completely decoupled formulation of the effective stress response from damage. As alluded to in the sequel (8)–(10) the damage evolution is defined as a postprocessing of the effective stress response. Concerning the damage evolution $\dot{\phi} \geq 0$ we consider developments in, for example, Chuzhoy et al. [4] and Cocks and Ashby [20] and propose

$$\dot{\phi} = (1-\phi)\left((1-\phi)^{-(n_d+1)} - (1-\delta)\right)\beta[r]\lambda \geq 0, \tag{11}$$

where we introduced a slight adjustment to its original expression in terms of the parameter δ (normally set to $\delta = 1\%$), is introduced in order to define the effect of materials integrity in the beginning of the damage process. Furthermore, n_d is the damage exponent parameter for the material.

Another adjustment of the original damage evolution law refers to the β function which is the stress triaxiality factor defined by

$$\beta = \begin{cases} \sinh\left[\dfrac{2(2n_d-1)}{2n_d+1}r\right] & r \geq 0.8 \\ k_1 \exp[k_2 r] & r < 0.8, \end{cases} \tag{12}$$

where $r = -\bar{p}/\bar{\tau}_e$ and $\bar{p} = -(1/3)\mathbf{1}:\bar{\tau} = -(1/3)\mathbf{1}:\bar{\tau}^{tr}$. An enhancement of the original β-function in (12) is made to avoid unphysical negative β-values. This occurs for triaxiality ratios $r \leq 0.8$, where a new exponential branch to the hyperbolic function is defined. The parameters k_1 and k_2 are determined at the point r=0.8 so that.

$$\sinh\left[\frac{2(2n_d-1)}{2n_d+1}r\right] = k_1\exp[k_2 r],$$

$$\sinh\left[\frac{2(2n_d-1)}{2n_d+1}r\right]' = k_1 k_2 \exp[k_2 r]. \tag{13}$$

An Uncoupled Damage Model

Parallel to the continuum damage model outlined in the previous subsection we shall consider the uncoupled damage model of Johnson and Cook [14]. A "damage" D measure, represented by the accumulated plastic deformation at the "current" time t, is introduced as

$$D = \int_0^t \frac{\lambda}{\epsilon_f^p[t]} dt. \tag{14}$$

Whenever D reaches the value 1 at any one integration point in an element the element is removed from the mesh following the procedure of the element removal technique used in the analysis. For the review of the element removal method compare, for example, [21]. According to the Johnson-Cook model, the fracture strain is expressed by three dependencies in a multiplicative fashion (like in the JC material model). The dependencies are those of stress triaxiality, strain rate, and temperature formulated in terms of the equivalent plastic fracture strain ε_f^p defined as

$$\epsilon_f^p = (d_1 + d_2 \exp[-d_3 r])\left(1 + d_4 \ln\left[\frac{\lambda}{\dot{\epsilon}_0}\right]\right)(1 + d_5 \hat{\theta}). \tag{15}$$

The presence of hydrostatic tension significantly decreases the level of critical plastic strain at which the material is considered to fracture. This is because nucleation, growth and coalescence of voids (being the major driving force of ductile fracture) are generally promoted by the hydrostatic tensile stress. The five material parameters in the failure criterion are the initial failure strain d_1, the exponential factor d_2, the triaxiality factor d_3, the strain rate factor d_4, and the temperature factor d_5. Although the influence on the stress response is different for the uncoupled and continuum damage models, the development of the "damage" variable is controlled by the triaxiality ratio r, the strain rate, and the temperature in both cases. We note that the same dependencies in both failure models open up for the possibility to fit the associated damage parameters to obtain a calibrated similar response from both damage models.

MESH DEPENDENCE INVESTIGATION

It is well known that local damage models generally lead to a pathological mesh dependence in the FE representation of localized plastic deformation. It is, however, argued in the literature (cf., e.g., [17]) that a viscous regularization of the continuum material model, for example, via viscoplasticity, may act as a localization limiter. To investigate the significance of this statement, the two (rate dependent) damage representations, described in the previous section, will be considered in the mesh dependency investigation.

The modeling of the effective material stress response (that serves as a basis for both the uncoupled and the continuum damage calculations) is based on the hypoelastic inelastic framework applied to the JC plasticity model [1]. Although the models are phrased in the thermodynamically consistent hyperelasticity framework, the hypoelastic inelastic framework is chosen due to its computational efficiency, as discussed in [13]. The hypoelastic inelastic response is postulated as

$$\hat{\bar{\tau}} = \mathbf{E}_e : \mathbf{l}_e \text{ with } \mathbf{E}_e = 2G\mathbf{I}_{\text{dev}}^{\text{sym}} + K\mathbf{1} \otimes \mathbf{1} \qquad (16)$$

where $\bar{\tau}$ is the effective Green-Naghdi stress rate.

Matching the Parameters between Uncoupled Damage and Continuum Damage Formulations

In the previous subsection we outlined two fundamentally different damage/failure models. In the first coupled one, the damage is evolving progressively with the plastic deformation, whereas in the uncoupled one the evolved damage is assumed to take place in a single step, as discussed in Section 2.4. However, we note that the total stress response of the two models is comparable for fairly high values of the damage parameter (n_d), corresponding to rapid damage evolution when $\phi \to 1$. It appears that the value $n_d = 16$ is reasonable from both experimental and numerical investigations of pearlitic steel at room temperature (cf. [5]). In this case, the parameters of the JC-failure model are calibrated to the parameters of the continuum damage model. To this end, the parameters d_1–d_4 of the uncoupled model are determined via least squares fitting from four loading cases. For this purpose, the uniaxial stress and simple shear tests at the material point level were considered with two different loading rates. The resulting stress-strain behavior after the calibration is shown in Figure 1, for each loading case and damage model. The nice fit between the models is noteworthy. The resulting parameters (d_1–d_4) are shown in Table 3.

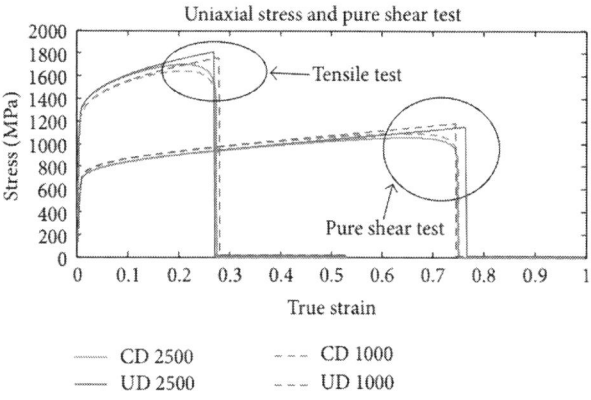

Figure 1: Resulting stress-strain behavior after the calibration of the CD and UD models based on uniaxial stress and pure shear deformation test with the applied loading rates 1000 and 2500.

The values of Young's modulus, Poisson's ratio, and density for the pearlite used in the simulations have been taken from the literature (cf. [4, 22] and cf. also Table 1 for representative values). The pearlite material parameters for the JC plasticity model (Table 2) are taken from [13] where the parameters are calibrated based on experimental data for pearlitic rail steels reported in [23, 24].

Table 1: Material properties pertinent to elastic and thermal response

E (GPa)	v	ρ (kg/m³)	T_{melt} (K)
190	0.3	7850	1673

Table 2: Johnson-Cook parameters

A (MPa)	B (MPa)	c	n	m	$\dot{\varepsilon}_0$
550	500	0.0804	0.4	1.68	0.001

Table 3: Calibrated uncoupled damage model parameters

d_1	d_2	d_3	d_4	d_5
0.26	0.614	2.557	-0.028	0.6

FE Analyses of a Shear Loaded Pearlite Plate

The mesh dependency investigation is performed both for the continuum damage model and for the uncoupled damage model. These two models have been used along with the hypoelastic inelastic framework applied to the JC-plasticity model and they have, for this purpose, been implemented in the commercial software Abaqus/Explicit as separate user subroutines.

The investigation is based on the results from the simulation of a shear loaded 2D-plate with (discussed earlier) pearlite material properties. The FE representations of the plate created are square shaped with dimensions of 50 × 50 mm. Different displacement rates were used

(500, 2500, and 10000 mm/s) in the analyses. The calculations are performed isothermally in order to focus on the damage effect.

To be able to perform the mesh dependency investigation, four FE representations of the plate are created with four different sizes of the four-node plain strain bilinear elements and with approximate element sizes of 0.1, 0.3, 0.5, and 1.0 mm; the models contain 69921, 33899, 11933, and 2931 elements, respectively. The different FE representations of the plate are presented in Figure 2. The meshes (a)–(c) have uniform element size distributions, whereas mesh (d) of the plate has a finer element size in the mid area. Free distribution of the element nodes is used during creation of the meshes to avoid possible effects of structured mesh patterns.

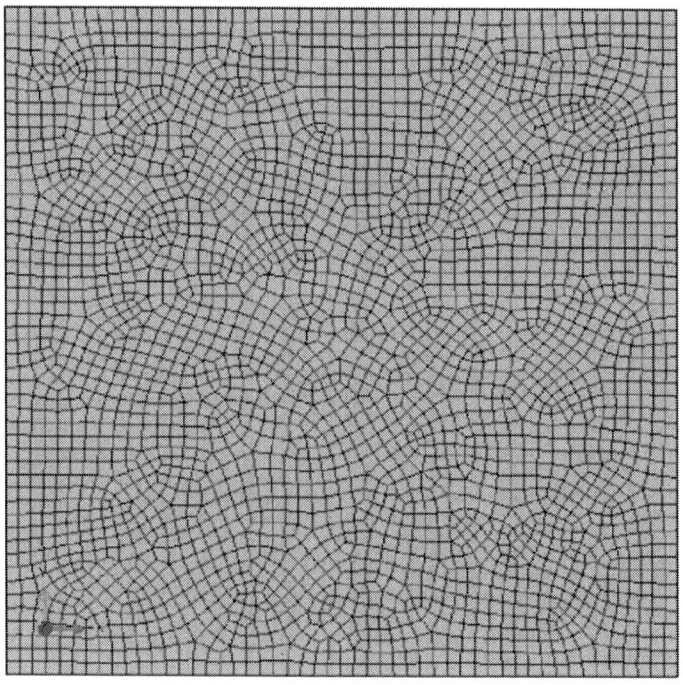

(a)

Rate Sensitive Continuum Damage Models and Mesh Dependence...

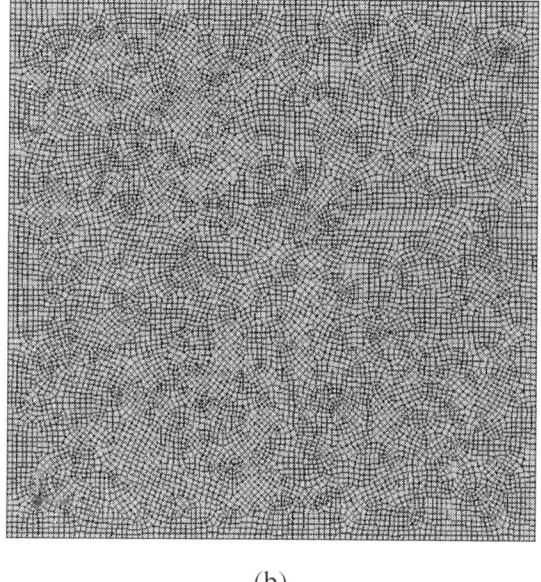

(b)

(c)

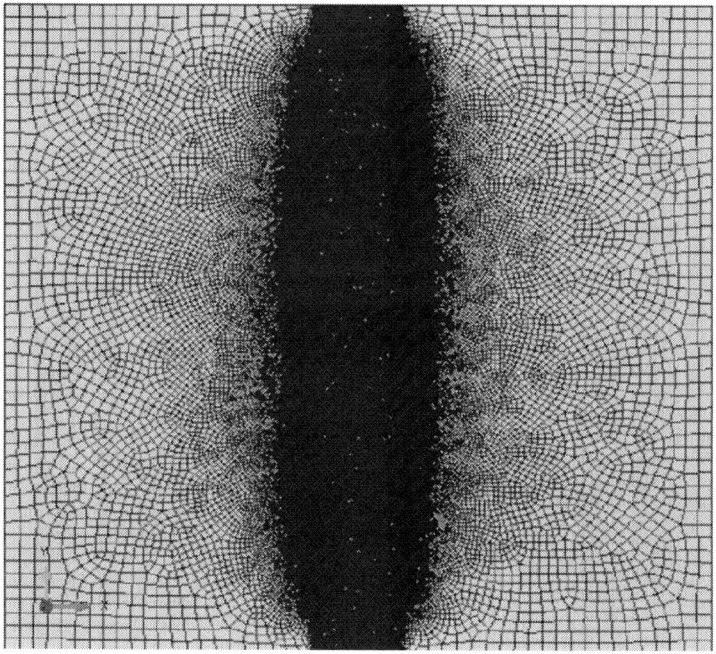

(d)

Figure 2: The considered FE representations of the plate based on different element sizes (1.0, 0.5, 0.3, and 0.1 mm corresponding to (a), (b), (c), and (d)). The meshes (a)–(c) are referred to as "coarse," "mean," and "fine," whereas mesh (d) is referred to "extra fine."

The set-up for the shear test is shown in Figure 3. The mid area of the plane strain plate (highlighted in Figure3) is exposed to severe shear deformation by prescribed vertical displacement (12.5 mm) downwards of the nodes on the left side of the top edge of the model. The nodes on the other side (right) of the localization area, lying both on the top and bottom edge, are constrained in the vertical direction. Also, the nodes on the left and the right plate edges are constrained in the horizontal direction.

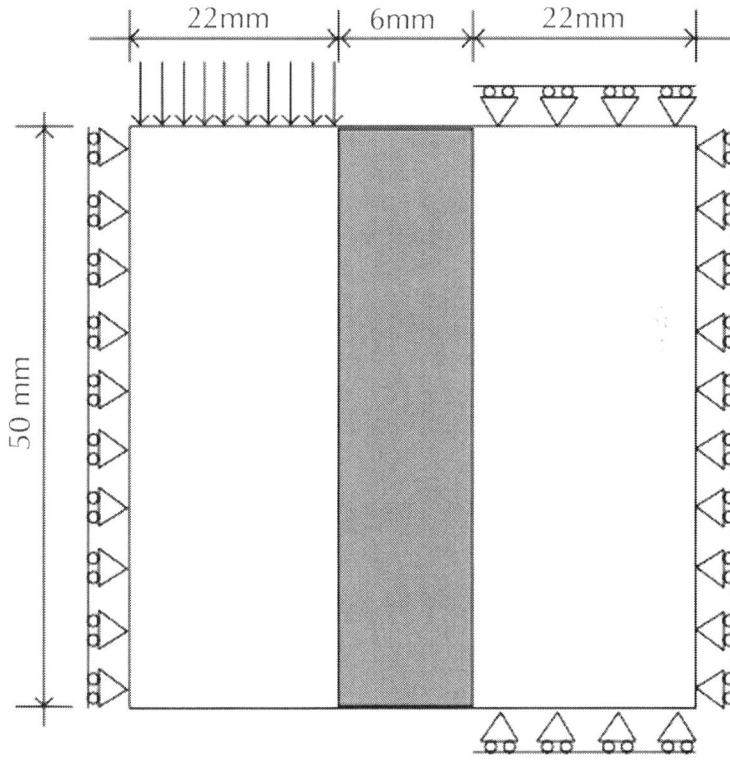

Figure 3: Considered shear loaded plate in plane strain.

RESULTS

In this section, the simulation results in terms of shear deformations of different FE representations of the plate are compared and analyzed with respect to mesh dependence. Simulations are conducted based on both the uncoupled and the continuum damage models, with different displacement rates in order to reveal possible convergence with respect to mesh dependency due to viscous regularization. The study is conducted based on results in the form of the force-displacement curves, where the magnitude of the force is calculated at the edge where the displacement boundary conditions are defined.

We emphasize that the variation of the crack patterns and deformations obtained for different cases, where the involved damage models, displacements rates, and FE representations are varied, must also be taken into account when analyzing the results. In the simulations, where all four FE representations are used, the cracks are developing through one element row irrespectively of element size, which is typical for a pathological mesh dependency behavior. The overall impression is that simulations based on the uncoupled damage model generally behave more consistently regarding deformations and crack patterns obtained with different FE representations and displacement rates.

The results of the simulations, in form of force-displacement curves, where both damage models and all element sizes were used, are plotted together in Figures 4(a), 4(b), and 4(c). As can be observed from the results, the damage development for the two damage models occurs almost simultaneously. This shows that the uncoupled and continuum damage models behave in a similar fashion in the shear test simulations, after the calibration of parameters of the uncoupled damage against the continuum damage model. If a comparison is made between the force-displacement results for different FE representations, a clear mesh dependence appears with a typical reduction of dissipated energy upon mesh refinement. This is despite the fact that a rate dependent model is utilized. This is illustrated, for instance, in the case of the continuum damage model (continuous lines) with displacement rate 2500 mm/s shown in Figure 4(b). In the case of the uncoupled damage model, the trend is the same until later in the damage development process where the curves (dashed lines) are crossing each other and the opposite trend is obtained which leads to smaller differences in energy dissipation compared to results based on the continuum damage model. If the crack patterns for the different displacement rates are compared, the only significant variation observed is in the case of the fine mesh (Figure2(c)) for the simulations with the continuum damage model. Here, the location of the initiation point of the crack on the upper right edge of the plate moves to the right as the displacement rate is increased (cf Figure5(a)).

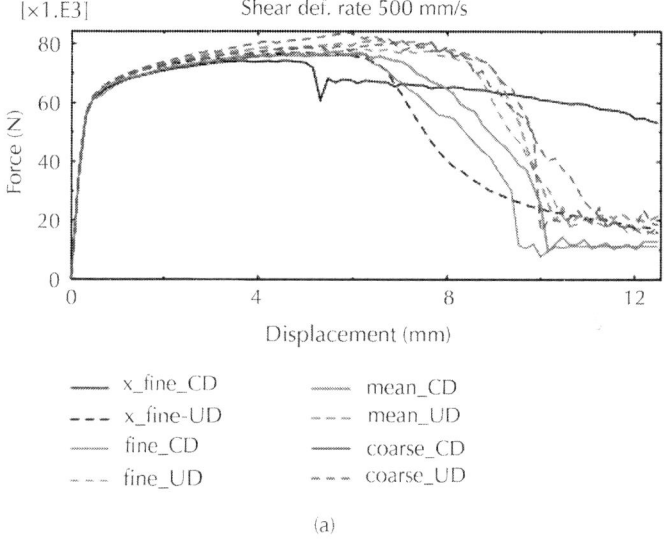

(a)

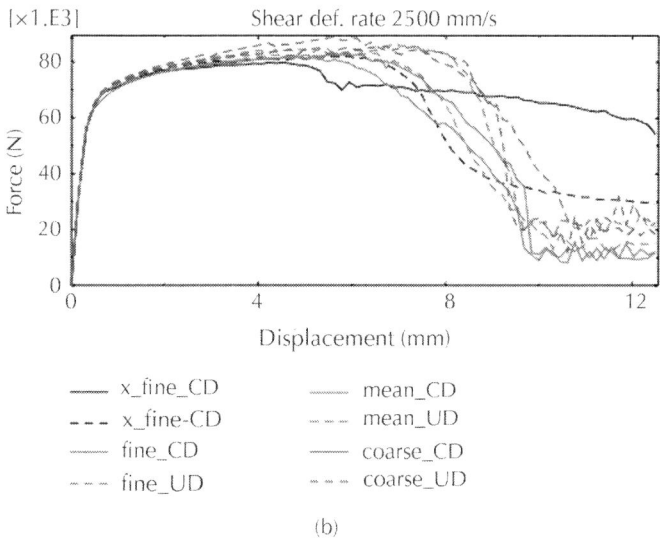

(b)

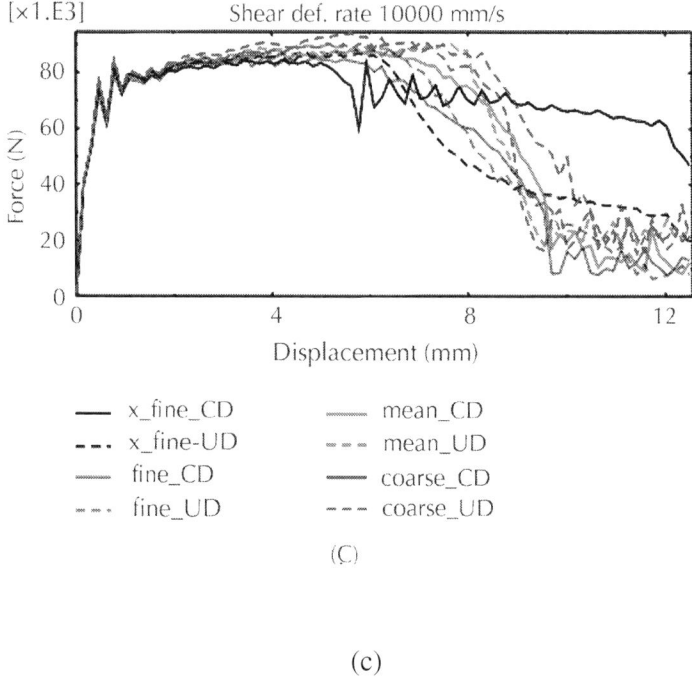

(c)

Figure 4: Force [N]-displacement [mm] curves from shear test analyses with displacement rates: (a) 500 mm/s, (b) 2500 mm/s, and (c) 10000 mm/s.

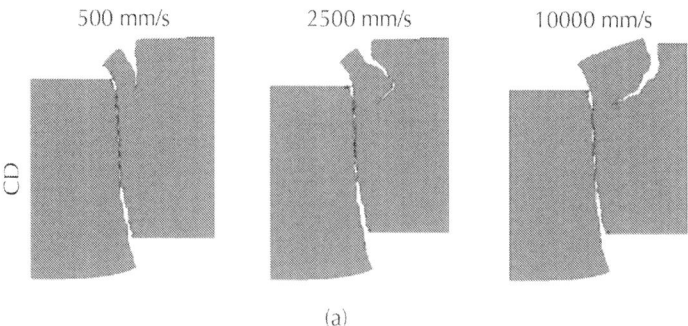

(a)

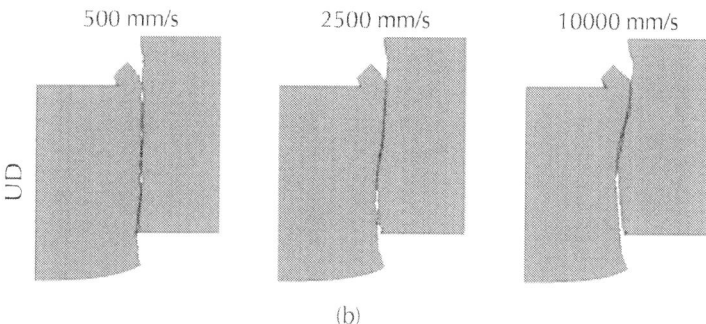

(b)

Figure 5: Variation of crack patterns in fine mesh for different displacement rates for (a) continuum damage model (CD) and (b) uncoupled damage model (UD).

A possible regularizing effect of higher deformation rates on the mesh dependency is also investigated by comparison of the inherent distances between force-displacement curves belonging to different displacement rates. Only a minor difference is observed and it was not possible to draw any general conclusions. As expected, in the simulations with displacement rate 10,000 mm/s the curves are becoming more uneven (due to dynamical effects) as compared to the lower rates. The black curves (named "x_fine" in the legend) represent values belonging to the finest mesh depicted in Figure 2(d). Their deformation patterns follow different trends compared to the rest of the curves, in particular, when the continuum damage model is used.

CONCLUDING REMARKS

One way of enhancing the implementation and computational efficiency of the simulation process is to use a decoupled formulation of the effective stress response from damage. Thereby, a common effective stress routine that serves as a base for the both failure models is used in present work.

A general conclusion from the simulation results presented in this paper is that a clear mesh dependency is present in all deformation modes and for all deformation rates, both for the uncoupled and for the coupled damage model considered in the investigation. The set of element sizes, deformation rates, and material parameters in the present investigation, relevant from the engineering point of view, appears not to yield any regularizing effect on the mesh dependency in the great majority of the cases (possibly all), despite the employment of the rate dependent damage models. As argued in the literature [16], one explanation is that, under the current "engineering" circumstances, the viscous regularization effect is too low. Furthermore, the effective viscoplastic regularization length decreases with the increasing damage (or deformation) values and when it reaches the characteristic element size, the regularization effect disappears. Consequently, a larger difference between the effective viscoplastic length scale and element size increases the regularization effect and could be accomplished by increase of viscosity, increase of the deformation rates, and possibly also further refinement of the mesh.

If the comparison is made between the continuum and uncoupled damage models in the shear test, a good agreement is obtained between the resulting force-displacement curves. The obtained differences between results based on the two types of damage models in terms of (in some cases) different crack patterns and energy needed to initiate the cracks, despite the adopted calibration procedure of the model parameters in order to obtain analogous behavior, possibly depend on the different nature of the damage models impact on the continuum behavior. In order to evaluate which failure model (and associated set of parameters) yields the most reliable results, physical experiments are needed. In general, it seems like, despite its restrictions concerning lack of damage influence on the continuum behavior, the uncoupled damage model gives more stable and realistic results, regarding crack development, compared to the present continuum damage model.

REFERENCES

1. G. R. Johnson and W. H. Cook, "A constitutive model and data for metals subjected to large strains, high strain rates and high temperatures," in Proceedings 7th International Symposium on Ballistics, pp. 541–547, 1983.
2. D. J. Bammann, "Modeling temperature and strain rate dependent large deformation of metals," Applied Mechanics Reviews, vol. 43, no. 5, pp. S312–S319, 1990.

3. D. J. Bammann, M. L. Chiesa, and G. C. Johnson, "Modeling large deformation and failure in manufacturing process," in Theoretical and Applied Mechanics, T. Tatsumi, E. Wannabe, and T. Kambe, Eds., pp. 359–376, 1996.
4. L. Chuzhoy, R. E. DeVor, S. G. Kapoor, and D. J. Bammann, "Microstructure-level modeling of ductile iron machining," Journal of Manufacturing Science and Engineering, vol. 124, no. 2, pp. 162–169, 2002.
5. L. Chuzhoy, R. E. DeVor, S. G. Kapoor, A. J. Beaudoin, and D. J. Bammann, "Machining simulation of ductile iron and its constituents, part 1: estimation of material model parameters and their validation," Journal of Manufacturing Science and Engineering, vol. 125, no. 2, pp. 181–191, 2003.
6. S. L. Soo and D. K. Aspinwall, "Developments in modelling of metal cutting processes," Proceedings of the Institution of Mechanical Engineers L: Journal of Materials: Design and Applications, vol. 221, no. 4, pp. 197–211, 2007.
7. T. D. Marusich and M. Ortiz, "Modelling and simulation of high-speed machining," International Journal for Numerical Methods in Engineering, vol. 38, no. 21, pp. 3675–3694, 1995. · ·View at Zentralblatt MATH
8. E. Ceretti, P. Fallböhmer, W. T. Wu, and T. Altan, "Application of 2D FEM to chip formation in orthogonal cutting," Journal of Materials Processing Technology, vol. 59, no. 1-2, pp. 169–180, 1996.
9. T. D. Marusich and M. Ortiz, "Simulation of chip formation in high-speed machining," in Proceedings of the Joint ASME Applied Mechanics and Materials Summer Meeting, pp. 127–139, June 1995.
10. T. Obikawa, H. Sasahara, T. Shirakashi, and E. Usui, "Application of computational machining method to discontinuous chip formation," Transactions of the ASME: Journal of Manufacturing Science and Engineering, vol. 119, no. 4, pp. 667–674, 1997.
11. E. Ceretti, M. Lucchi, and T. Altan, "FEM simulation of orthogonal cutting: serrated chip formation," Journal of Materials Processing Technology, vol. 95, no. 1–3, pp. 17–26, 1999.

12. G. Ljustina, R. Larsson, and M. Fagerström, "A FE based machining simulation methodology accounting for cast iron microstructure," Finite Elements in Analysis and Design, vol. 80, pp. 1–10, 2014.
13. G. Ljustina, M. Fagerström, and R. Larsson, "Hypo- and hyper-inelasticity applied to modeling of compacted graphite iron machining simulations," European Journal of Mechanics A: Solids, vol. 37, pp. 57–68, 2013.
14. G. R. Johnson and W. H. Cook, "Fracture characteristics of three metals subjected to various strains, strain rates, temperatures and pressures," Engineering Fracture Mechanics, vol. 21, no. 1, pp. 31–48, 1985.
15. Abaqus Analysis User›s Manual, Version 6.10, Providence, RI, USA, 2010.
16. M. S. Niazi, H. H. Wisselink, and T. Meinders, "Viscoplastic regularization of local damage models: Revisited," Computational Mechanics, vol. 51, no. 2, pp. 203–216, 2013.
17. V. Dias da Silva, "A simple model for viscous regularization of elasto-plastic constitutive laws with softening," Communications in Numerical Methods in Engineering, vol. 20, no. 7, pp. 547–568, 2004.
18. A. Needleman, "Material rate dependence and mesh sensitivity in localization problems," Computer Methods in Applied Mechanics and Engineering, vol. 67, no. 1, pp. 69–85, 1988.
19. R. de Borst, L. J. Sluys, H.-B. Muhlhaus, and J. Pamin, "Fundamental issues in finite element analyses of localization of deformation," Engineering Computations, vol. 10, no. 2, pp. 99–121, 1993.
20. A. C. F. Cocks and M. F. Ashby, "Intergranular fracture during power-law creep under multiaxial stresses," Metal science, vol. 14, no. 8-9, pp. 395–402, 1980.
21. J.-H. Song, H. Wang, and T. Belytschko, "A comparative study on finite element methods for dynamic fracture," Computational Mechanics, vol. 42, no. 2, pp. 239–250, 2008.
22. N. K. Fukumasu, P. L. Pelegrino, G. Cueva, R. M. Souza, and A. Sinatora, "Numerical analysis of the stresses developed during the sliding of a cylinder over compact graphite iron," Wear, vol. 259, no. 7-12, pp. 1400–1407, 2005.

23. J. Ahlström and B. Karlsson, "Fatigue behaviour of rail steel—a comparison between strain and stress controlled loading," Wear, vol. 258, no. 7-8, pp. 1187–1193, 2005.
24. M. S. J. Hashmi and M. N. Islam, "Stress—strain properties of railway steel at strain rates of up to 105 per second," in Transactions of the International Confer

Chapter 8

Nano-scale Machining of PolycrystallineCoppers - Effects of Grain Size and Machining Parameters

Jing Shi[1], Yachao Wang[1], and Xiaoping Yang[2]

[1]Department of Industrial and Manufacturing Engineering, North Dakota State University, Fargo, ND 58108, USA

[2]EBU Global Manufacturing Engineering Department, Cummins Inc., Columbus, IN 47202, USA

ABSTRACT

In this study, a comprehensive investigation on nano-scale machining of polycrystalline copper structures is carried out by molecular dynamics (MD) simulation. Simulation cases are constructed to study the impacts of grain size, as well as various machining parameters. Six polycrystalline copper structures are produced, which have the corresponding equivalent grain sizes of 5.32, 6.70, 8.44, 13.40, 14.75, and 16.88 nm, respectively. Three levels of depth of cut, machining speed, and tool rake angle are also considered. The results show that greater cutting forces are required in nano-scale polycrystalline machining with the increase of depth of cut, machining speed, and the use of the negative tool rake angles. The distributions of equivalent stress are consistent with the cutting force trends. Moreover, it is discovered that in the grain size range of 5.32 to 14.75 nm, the cutting forces and equivalent stress increase with the increase of grain size for the nano-structured copper, while the trends reserve after the grain size becomes even higher. This discovery confirms the existence of both the regular Hall–Petch relation and the inverse Hall–Petch relation in polycrystalline machining, and the existence of a threshold grain size allows one of the two relations to become dominant. The dislocation-grain boundary interaction shows that the resistance of the grain boundary to dislocation movement is the fundamental mechanism of the Hall–Petch relation, while grain boundary diffusion and movement is the reason of the inverse Hall–Petch relation.

BACKGROUND

Built on the classical Newton's Second Law, molecular dynamics (MD) simulation has been proven to be an effective tool to study many underlying intriguing mechanisms of material processing. This technique works particularly well with very small scales, which could be often ineffective for any experimental approaches or other mainstream numerical simulation approaches. As such, it has been applied to tackle countless interesting problems in the area of material processing, including the formation of dislocation, development of fracture, evolution of friction and wear, and effects of processing parameters in various processes. Nano-scale machining is one of those processes, and it is an important method to create miniaturized components and features. A substantial amount of research has been carried out on nano-scale machining by MD simulation. The pioneer works of Inamura et al. [1,2] adopted this technique to investigate the mechanics, energy dissipation, and shear deformation in nano-scale machining of monocrystal copper. It was argued that the theory of continuum mechanics is not applicable to nano-scale machining. Meanwhile, the deformation mechanism in the primary shear zone seems to be related to buckling due to severe compression in that area, while the deformation at the secondary shear zone appears to be the result of shear plastic deformation by yield shear stress.

Numerous other studies on MD simulation of nano-scale machining have emerged since 1990s. Ikawa et al. [3] investigated the minimum thickness of cut (MTC) for ultrahigh machining accuracy. It was discovered that an undercut layer of 1 nm is achievable for machining of monocrystal copper with a diamond tool. Fang and Weng [4] also simulated nano-scale machining of monocrystal copper using a diamond tool by focusing on friction. It was found that the calculated coefficients of friction in nano-scale machining are close to the values obtained in macro-scale machining. Shimada et al. [5, 6] adopted MD simulation to analyze 2D machining of monocrystal copper using diamond tools. It was found that disordered copper atoms due to tool/material interaction can be self re-arranged after the cutting edge passes the affected area. For simulating nano-scale machining of monocrystal copper, Ye et al. employed the embedded atom method (EAM) to model the potential energy of copper atoms [7]. Compared with other potential energy models for nano-scale machining, the EAM potential can produce comparable results, and thus, it is regarded as a viable alternative. Komanduri et al. [8, 9] conducted extensive simulation works on nano-scale machining of monocrystal aluminum and silicon. The works reveal the effects of various parameters, such as cutting speed, depth of cut, width of cut, crystal orientation, and rake angle, on chip formation and cutting force development. The effort on investigating the effects of machining parameters on the performances of nano-scale machining has never stopped. For instance, Promyoo et al. [10] investigated the effects of tool rake angle and depth of cut in nano-scale machining of monocrystal copper. It was discovered that the ratio of thrust force to tangential cutting force decreases with the increase of rake angle, but it hardly changes with the depth of cut. Shi et al. [11] developed a realistic geometric configuration of three-dimensional (3D) single-point turning process of monocrystal copper and simulated the creation of a machined surface based on multiple groove cutting. A variety of machining parameters were included in this realistic 3D simulation setting. Meanwhile, other phenomena in nano-scale machining are also investigated by MD simulation approach. Tool wear appears to be one of the most studied topics. Zhang and Tanaka [12] confirmed the existence of four regimes of deformation in machining at atomistic scale, namely, no-wear regime, adhering regime, ploughing regime, and cutting regime. It was found that a smaller tip radius or a smaller sliding speed brings a greater no-wear regime. Cheng et al. [13] discovered that the wear of a diamond tool is affected by the cutting temperature as heat generation decreases the cohesive energy between carbon atoms. Another study of nano-machining revealed that the iron workpiece has anisotropic influence on diamond cutting tool graphitization, an important indicator of tool wear [14]. In addition, there are a number of studies in the literature on the brittle-ductile transition phenomenon of silicon material in nano-scale machining or indentation. For instance, Tanaka et al. observed amorphous phase transformation of silicon in nano-machining and that stable shearing of the amorphous region is necessary for ductile-mode machining [15]. Also, a numerical study of surface residual stress distribution of silicon during nano-machining process is presented by Wang et al. [16]. Their MD simulation results revealed that higher hydrostatic pressure beneath the tool rake face induces more drastic phase transformation and thus generates more compressive surface residual stress. MD simulation is also capable of modeling chip formation, separation, and evolution mechanism. For instance, Ji et al. [17] studied the tool-chip stress distribution in nano-machining of copper, and the results were compared to the existing models of conventional machining. Lin and Huang [18] studied nano-cutting process by MD simulation and proposed the innovative 'combined Morse potential function and rigid tool space restrictions criterion' as the chip separation criterion. It was used to establish the shape function of the FEM-MD combined model.

Existing studies on MD simulation of nano-scale machining usually adopt defect-free monocrystalline structures as the work material [19]. The most popular ones have been monocrystal copper, aluminum, and silicon. Nevertheless, the vast majority of engineering materials exist in polycrystalline (instead of monocrystalline) forms. It is not difficult to understand that machining polycrystalline structures may yield different results compared with machining monocrystalline structures. Moreover, the grain size in polycrystalline structures is often a controlling factor for material properties and material responses to deformation. It is important to investigate how it impacts the machining performance at nano/atomistic scale. In a preliminary study, Shi and Verma [20] constructed one polycrystalline copper structure, simulated nano-scale machining of the structure, and made a comparison with monocrystalline machining. It was discovered that for all cutting conditions simulated, the polycrystalline structure requires smaller cutting forces compared with the monocrystalline structure. This result might be expected as the existence of grain boundary is usually regarded as defects, and thus, it reduces material strength. However, many more interesting questions arise from the preliminary finding, such as 'Will the polycrystalline structures with different grain sizes behave differently in nano-scale machining?', 'What are the roles of grain boundaries in measured machining performances?', and 'How do the machining parameters affect the performances of nano-scale polycrystalline machining?' In an effort to answer these research questions, we carry out this study.

The effect of grain size and grain boundary on the material's mechanical property has been well discussed. Usually, the well-known Hall–Petch relationship is widely accepted. This relationship indicates that material strength increases with the decrease of grain size. However, for very fine nano-structured materials, this relationship may no longer hold. Yang and Vehoff [21] investigated the dependency of hardness upon grain size in nano-indentation experiments. With the indentation depth of less than 100 nm, it is clearly revealed that the local interaction between dislocations and grain boundaries causes various hardness dependences on indentation depth. Zhang et al. [22] carried out nano-indentation experiments on copper with grain sizes from 10 to 200 nm. It was found that at short dwell times, the hardness increases significantly with decreasing grain size. However, the difference substantially diminishes at longer times due to the rapid grain growth under the indenter. Similar reverse proportion relations between grain size and hardness are observed in indentation experiments at micro-scale in the literature. Li and Reece [23] discovered that grain size has a significant effect on surface fatigue behavior, and increasing grain size reduces the threshold for crack nucleation. Also, Lim and Chaudhri [24] showed that in the grain size range of 15 to 520 μm, the initial higher dislocation density for smaller grains is believed to cause higher Vickers hardness. More importantly, the rapid advance of numerical simulation techniques has enabled more detailed analysis of dislocations and grain boundaries in deformation of polycrystallines. For instance, with the help of MD simulation, the interaction of dislocations with a $\Sigma=5(210)$ [001] grain boundary is analyzed, and the transmission of dislocation across the grain boundary is observed [25]. Another MD simulation study indicates that compared to bulk diamond crystal, substitution energies are found to be significantly lower for grain boundaries [26].

The remainder of the paper is organized as follows. In the next section, the MD model construction for nano-scale machining of polycrystalline is briefly introduced. The machining conditions for the simulation cases are also summarized. Thereafter, the simulation results of nano-scale machining are presented, in which the major observations are made regarding the effects of grain size and machining parameters. More importantly, a detailed discussion on the grain size effect is provided to reveal the governing mechanism in nano-scale machining. Finally, conclusions are drawn and future research is pointed out in the last section.

METHODS

Simulation Model Construction

Figure 1 shows the overall MD simulation model constructed according to a 3D orthogonal machining configuration. For all the cases, the tool material is always diamond, and the work materials are polycrystalline coppers except for the benchmark case of monocrystalline copper. The diamond tool is oriented to achieve a rake angle of -30° and a relief angle of 30°, and it is treated as a rigid body in MD simulation. It can also be seen from Figure 1 that the work material atoms are categorized into three types - namely, fixed layer, thermostat layer, and Newton layer. The atoms in the fixed layer have fixed positions and only interact with the other two types of work material atoms. The thermostat layer lies between the fixed layer and the Newton layer. The atoms in the thermostat layer are used to stabilize the temperature of the system. For all the simulation cases, the copper workpieces have the identical dimension of $432 \times 216 \times 216$ Å^3. The polycrystalline copper structures are built based on the operation of Voronoi site-rotation and cut [27]. The simulation is carried out using LAMMPS, a general-purpose molecular dynamics simulation code developed by Sandia National Lab [28]. Post-processing codes are developed in-house to calculate machining forces, stress distributions, and dislocation development.

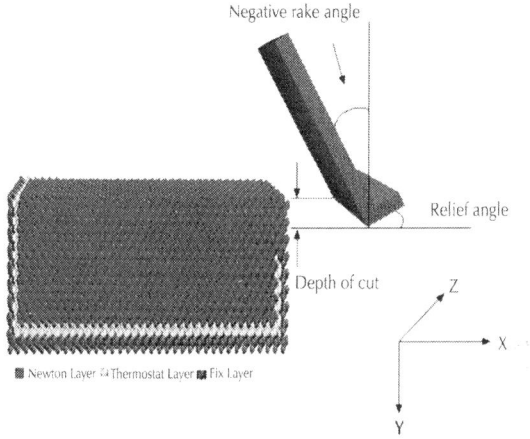

Figure 1: MD simulation model of nano-scale machining.

Simulated Machining Cases and Machining Parameters

A total of 13 simulation cases are constructed to investigate (1) the effects of machining parameters in polycrystalline machining and (2) the effect of grain size of polycrystalline copper on machining performances. Table 1 summarizes the machining conditions for all the 13 cases. For the first task, we select three levels of machining speed, i.e., 25, 100, and 400 m/s; three levels of depth of cut, i.e., 10, 15, and 20 Å; and three levels of tool rake angle, i.e., -30°, 0°, and +30°. As such, the group of cases C4, C8, and C9 can be used to investigate the machining speed effect since the only different parameter among the three cases is the machining speed. For the same reason, the group of cases C4, C10, and C11 can be used to reveal how the depth of cut affects polycrystalline machining, and cases C4, C12, and C13 can be compared to show the effect of tool rake angle. Note that the lowest machining speed employed in this study is 25 m/s, which is still high even compared with the typical machining speeds (e.g., 5 to 10 m/s) of high speed machining. However, this arrangement is necessary because MD simulation is extremely computation intensive. For instance, the average computation time for a case with 400 m/s machining speed in this study is about 8 days on an Intel Core i7 3.2-GHz PC.

Table 1: Machining conditions for the 13 simulation cases of nano-scale machining

Case number	Depth of cut (Å)	Tool rake angle (deg)	Machining speed (m/s)	Grain size (nm)
C1	15	-30	400	Monocrystal
C2	15	-30	400	16.88
C3	15	-30	400	14.75
C4	15	-30	400	13.40
C5	15	-30	400	8.44
C6	15	-30	400	6.70
C7	15	-30	400	5.32
C8	15	-30	100	13.40
C9	15	-30	25	13.40
C10	10	-30	400	13.40
C11	20	-30	400	13.40
C12	15	0	400	13.40
C13	15	30	400	13.40

Shi et al.

Shi et al. Nanoscale Research Letters 2013 8:500, doi:10.1186/1556-276X-8-500

For the second task, the machining conditions are fixed, and six levels of grain size are created. The six grain sizes are 5.32, 6.70, 8.44, 13.40, 14.75, and 16.88 nm. They correspond to 256, 128, 64, 16, 12, and 8 face-centered cubic (fcc) grains within an identical work dimension and represent simulation cases C2 to C7, respectively. The comparison among the six cases can illustrate the effect of grain size on polycrystalline machining. To make the comparison complete, a monocrystalline copper structure is also created and simulated, which is represented by case C1.

Potential Formulations

The interaction between the copper atoms in the work material and the carbon atoms in the diamond tool can be modeled using the pairwise Morse potential [29]:

$$U = D\{\exp[-2\alpha(r_{ij}-r_0)] - 2\exp[-\alpha(r_{ij}-r_0)]\}, \quad (1)$$

where D is the cohesion energy, α is a constant parameter, r_{ij} is the distance between the two atoms, and r_0 is the distance at equilibrium. The parameters for the Morse potential between copper and carbon atoms are presented in Table 2.

Table 2: Morse potential parameters for Cu-C interaction [1], [31]

Parameter	Value
D (eV)	0.1063
α (Å^{-1})	1.8071
r_0 (Å)	2.3386
Potential cutoff distance (Å)	6.5

Shi et al.

Shi et al. Nanoscale Research Letters 2013 8:500, doi:10.1186/1556-276X-8-500

The interaction forces between copper atoms are modeled using the EAM potential, which is a multi-body potential energy function in the following form [30]:

$$U = F_\alpha \sum_{j \neq i} \rho_i(R_{i,j}) + \frac{1}{2}\sum_{j \neq i} \phi_{\alpha\beta}(R_{i,j}), \quad (2)$$

where the total energy (U) on atom i is the sum of the embedding energy F and the short-range pair potential energy φ, ρ is the electron density, and α and β are the element types of atoms I and j. The embedding energy is the energy to put atom i in a host electron density (ρ_i) at the site of that atom. The pair potential term (φ) describes the electrostatic contributions. The EAM potential parameters are presented in Table 3.

Table 3: EAM potential parameters for Cu-Cu interaction [4], [20]

Parameter	Value
Lattice constant (Å)	3.62
Cohesive energy (eV)	-3.49
Bulk modulus (GPa)	137
C' (GPa)	23.7
C_{44} (GPa)	73.1
$\Delta(E_{bcc} - E_{fcc})$ (meV)	42.7
$\Delta(E_{hcc} - E_{fcc})$ (meV)	444.8
Stacking fault energy (mJ/m^2)	39.5
Vacancy: E_f (eV)	1.21

Shi et al.

Shi et al. Nanoscale Research Letters 2013 8:500, doi:10.1186/1556-276X-8-500

To calculate the cutting force, the individual interaction force on atom i due to atom j should be computed first by differentiating the potential energy. For each tool atom, the reaction forces should also be summed among its neighbor atoms. Then, the cutting force in vector form can be obtained by summing all the interaction forces on the cutting tool atoms:

$$F = \sum_{i=1}^{N_T} \sum_j \frac{\partial U(r_{ij})}{\partial r_{ij}}, \quad (3)$$

where F is the cutting force and NT is the number of atom in the cutting tool.

For the calculation of stress components s_{xx}, s_{yy}, s_{zz}, s_{xy}, s_{xz}, and s_{yz} of atom i, the following equation is used:

$$\chi = \frac{1}{\Omega} \sum_i^N \left(m_i v_i \otimes v_i + \frac{1}{2} \sum_{i \neq j} r_{ij} \otimes \frac{\partial U(r_{ij})}{\partial r_{ij}} \right) \quad (4)$$

where χ is the average virial stress component, Ω is the volume of the cutoff domain, m_i is the mass, v_i is the velocity of atom i, $\otimes$ denotes the tensor product of two vectors, and N is the total number atoms in the domain. To calculate the equivalent stress (S), the virial stress components are used:

$$S = \sqrt{3\left(s_{xy}^2 + s_{yz}^2 + s_{xz}^2\right) + \frac{1}{2}\left[\left(s_{xx}-s_{yy}\right)^2 + \left(s_{xx}-s_{zz}\right)^2 + \left(s_{zz}-s_{yy}\right)^2\right]}. \quad (5)$$

RESULTS AND DISCUSSION

Effect of Machining Parameters

Effect of Depth of Cut

As mentioned above, cases C10, C4, and C11 adopt three levels of depth of cut, namely, 10, 15, and 20 Å, respectively, while the other machining parameters are the same. For each case, three snapshots of machining progress at the tool travel distances of 30, 120, and 240 Å are presented. The results for the three cases are shown in Figures 2, 3, and 4, respectively. First of all, chip formation progress can be observed here. For all the three cases, the machined chip accumulates in front of the tool rake face as the tool advances. The chip volume is approximately proportional to the depth of cut. However, the cutting chip thicknesses for cases C10, C4, and C11 are measured to be 18, 40, and 45 Å, respectively. The increase of chip thickness is more significant when the depth of cut increases from 10 to 15 Å, compared with the increase period from 15 to 20 Å.

232 Deformation and Fracture Mechanics of Engineering Materials

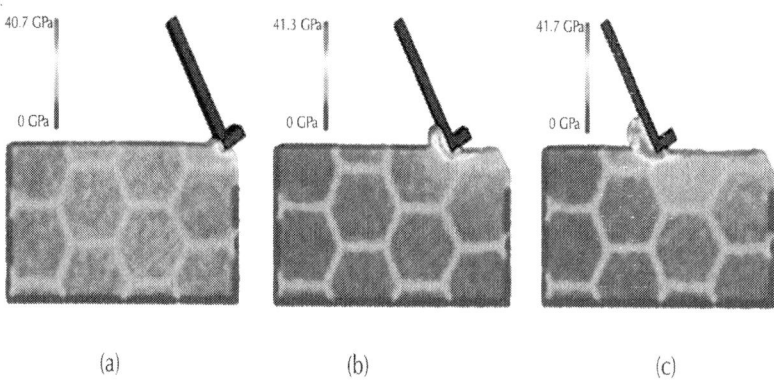

Figure 2: Chip formations and equivalent stress distributions in nano-scale polycrystalline machining for case C10. At the tool travel distances of (a) 30, (b) 120, and (c) 240 Å.

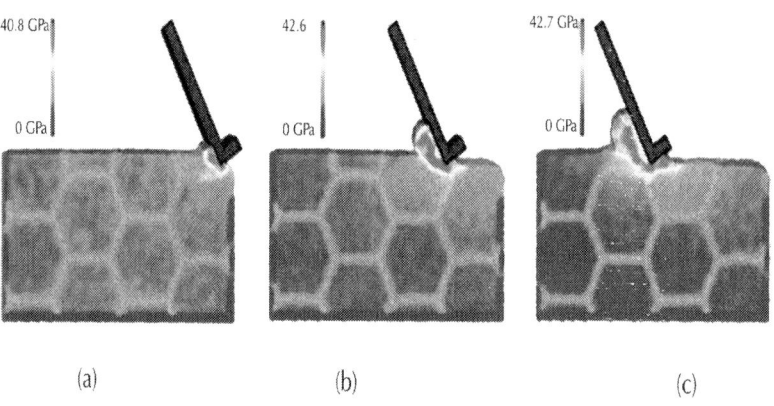

Figure 3: Chip formations and equivalent stress distributions in nano-scale polycrystalline machining for case C4. At the tool travel distances of (a) 30, (b) 120, and (c) 240 Å.

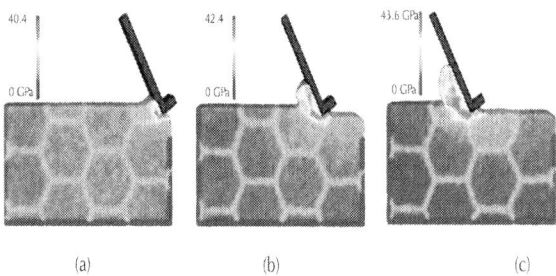

Figure 4: Chip formations and equivalent stress distributions in nano-scale polycrystalline machining for case C11. At the tool travel distances of (a) 30, (b) 120, and (c) 240 Å.

Figures 2, 3, and 4 also provide the information of equivalent stress distribution in polycrystalline machining. It can be found that the stress distribution pattern of nano-scale polycrystalline machining is overall consistent with that of conventional machining, as well as that of nano-scale machining of monocrystalline structures [20, 31]. For all the cases, the stress concentration is observed in the primary shear zone, where the chip is formed by high-strain-rate shearing in the primary shear zone, as well as the second shear zone, which is the friction-affected zone between the tool rake face and the chip. For each case, the maximum stress occurs at the primary shear zone and it increases as the depth of cut increases. For instance, at the tool travel distance of 240 Å, the maximum equivalent stress values are 41.7, 42.7, and 43.6 GPa for cases C10, C4, and C11, respectively. Meanwhile, our results indicate that the equivalent stress on grain boundaries is generally 30% to 60% higher than the stress inside the grains. Note that the difference of equivalent stresses on grain boundaries and inside the grains is not only caused by the exertion of cutting force. It is believed that the crystallographic orientation of grains could introduce stress concentration on and nearby boundaries. The literature also indicates that a higher amount of stress and lattice distortion can develop nearby the grain boundaries [32].

In addition, no crack is observed during the entire machining process for all cases. This is a reasonable result based on the MD simulation study by Heino et al. [33], in which the crack initiation and propagation mechanism within a copper material is investigated. It was estimated that the critical tensile stress for crack initiation is around 15 GPa. However, in our simulation, the maximum tensile stress of the as-machined surface in the vicinity of the cutting tool is around 3 GPa, which is much smaller than the critical crack initiation tensile stress. In addition, the use of a negative rake angle also helps avoid cracks and improve machined surface quality in nano-machining process [16].

Figure 5a,b compares the evolution curves of cutting force components, F_x and F_y, for cases C10, C4, and C11. F_x and F_y are the force components along the X and Y axes as indicated in Figure 1, and they represent the tangential force and the thrust force, respectively. It can be seen that for all the cases, both F_x and F_y increase rapidly at the beginning of machining process, but the trend of increase slows down after the tool travel distance is beyond about 30 Å. Overall, both the tangential and thrust forces increase with the increase of depth of cut. Nevertheless, a more significant increase in both force components is observed as the depth of cut increases from 10 to 15 Å, compared with that when the depth of cut increases from 15 to 20 Å.

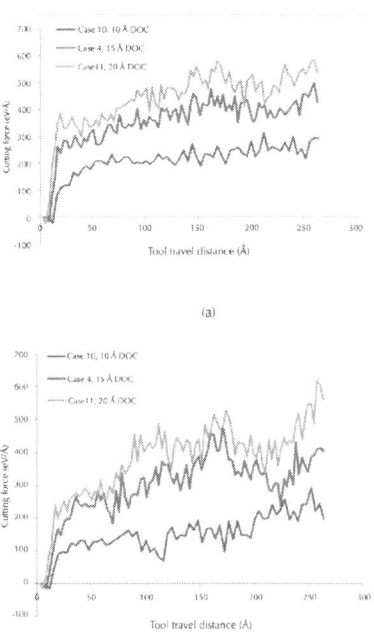

Figure 5: Evolution of cutting forces for three cases with three depths of cut (DOC). (a) Tangential force, F_x and (b) thrust force, F_y.

Meanwhile, to make a direct and fair comparison, the average F_x and F_y values are obtained by averaging the fluctuating force values obtained during the travel distance period of 160 to 280 Å, which represents the relative stable stage of the entire machining process. The results are summarized in Table 4. As the depth of cut increases from 10 to 15, and then to 20 Å, the tangential force increases from 254.41 to 412.16, and then to 425.32 eV/Å, and the thrust force increases from 199.99 to 353.59, and then to 407.26 eV/Å, respectively. The increase of cutting force due to the increase of depth of cut in nano-scale polycrystalline machining should not be a surprise. More energy is needed to remove more material, and this actually applies to the machining process at all length scales [10, 31, 34]. Moreover, the ratios of tangential force to thrust force, F_x/F_y, for the three cases are calculated. It is found that F_x/F_y decreases as the depth of cut increases. This means that as the depth of cut increases, the increase of thrust force is more significant than the increase of tangential force.

Table 4: Average cutting force values with respect to depth of cut

Case number	Depth of cut (Å)	F_x (eV/Å)	F_y (eV/Å)	F_x/F_y
C10	10	254.41	199.99	1.27
C4	15	412.16	353.59	1.17
C11	20	509.94	454.92	1.12

Shi et al.

Shi et al. Nanoscale Research Letters 2013 8:500, doi:10.1186/1556-276X-8-500

Effect of Tool Rake Angle

For this purpose, cases C4, C12, and C13 are compared because they adopt three different tool rake angles of -30°, 0°, and +30°, respectively. Figure 3 already shows the machining snapshots for case C4. Figures 6 and 7 illustrate the machining snapshots taken at the same tool travel distances for cases C12 and C13, respectively. Certainly, the rake angle dictates the chip formation/flow direction, and also, the chip geometries are somehow different among the three cases. By examining the equivalent stress distributions in the affected zones, it can be found that the primary shear zone becomes more distinguishable from the secondary shear zone when the rake angle changes from negative to positive. Also, the affected uncut zone ahead of the cutting tool becomes shallower when the rake angle changes from negative to positive. This indicates the severity of compression effect in the affected uncut zone.

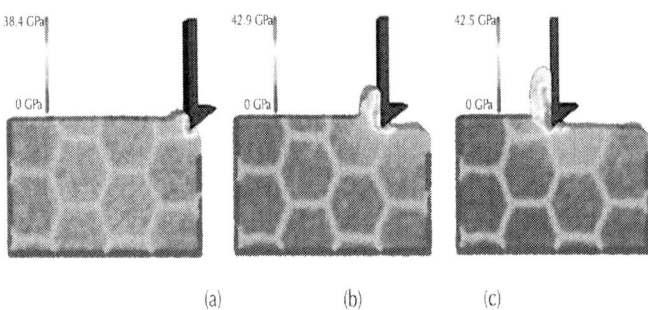

(a) (b) (c)

Figure 6: Chip formations and equivalent stress distributions in nanoscale polycrystalline machining for case C12. At the tool travel distances of (a) 30, (b) 120, and (c) 240 Å.

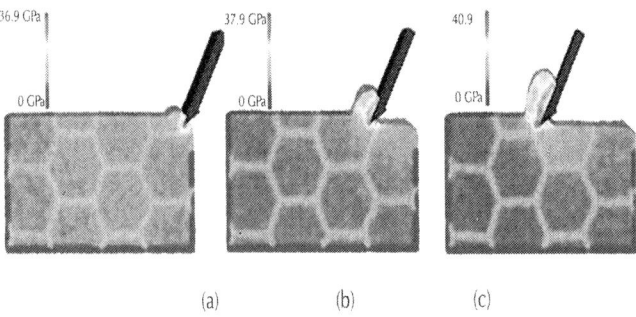

(a) (b) (c)

Figure 7: Chip formations and equivalent stress distributions in nanoscale polycrystalline machining for case C13. At the tool travel distances of (a) 30, (b) 120, and (c) 240 Å.

Similarly, the cutting force evolutions are compared to illustrate the effect of tool rake angle. As shown in Figure 8a,b, as the tool rake angle changes from -30° to 0°, and then to +30°, both the tangential force

F_x and the thrust force F_y decrease and the deduction of thrust force is more pronounced. The average F_x and F_y values are also calculated to make a more direct comparison. As shown in Table 5, with the -30°, 0°, and +30° tool rake angles, the average tangential forces are 412.16, 338.73, and 280.80 eV/Å, respectively, and the thrust force values are 353.59, 132.68, and 19.43 eV/Å, respectively. The ratio of tangential force to thrust force, F_x/F_y, increases from 1.17 to 14.45 as the rake angle changes from -30° to +30°. Clearly, the more drastic compression effect between tool and workpiece induced by the negative rake angle causes much higher thrust force compared to the cases with zero or positive tool rake angle. As the rake angle becomes more negative, the thrust force needs to increase more significantly compared to the tangential force to overcome the plastic deformation resistance of the work material under the tool tip. This result is consistent with the literature on conventional machining and nano-scale monocrystalline machining [35, 36].

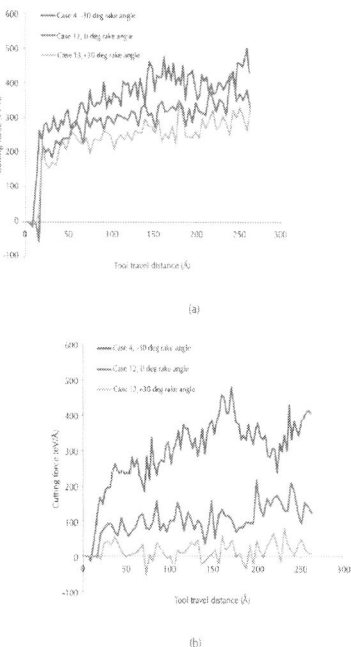

Figure 8: Evolution of cutting forces for three cases with three rake angles. (a) Tangential force, F_x and (b) thrust force, F_y.

Table 5: Average cutting force values with respect to tool rake angle

Case number	Tool rake angle (deg)	F_x (eV/Å)	F_y (eV/Å)	F_x/F_y
C4	-30	412.16	353.59	1.17
C12	0	338.73	132.68	2.55
C13	+30	280.80	19.43	14.45

Shi et al.

Shi et al. Nanoscale Research Letters 2013 8:500, doi:10.1186/1556-276X-8-500

Effect of Machining Speed

The effect of machining speed can be analyzed by comparing cases C4, C8, and C9, which employ the machining speeds of 400, 100, and 25 m/s, respectively. The chip formation and equivalent stress distribution for case C4 is already shown in Figure 3. Figures 9 and 10 depict the results of cases C8 and C9, respectively. The chip morphologies of cases C8 and C9 appear to be quite different from that of case C4. Under the lower machining speeds of 25 and 100 m/s, the chip formation is more like a material pile-up process, and the regular flow of the material along the tool rake face cannot be observed. Also, for these two lower speed cases, the stress concentration along the primary shear zone is more significant than that along the secondary shear zone. Therefore, chip formation seems to be very sensitive to the machining speed for nano-scale polycrystalline machining - the regular uniform chip can only be formed at high machining speeds of more than 100 m/s. In addition, it can be found that lower machining speeds reduce the maximum equivalent stress value. For instance, at the tool travel distance of 240 Å, the maximum equivalent stresses are 42.7, 31.2, and 30.1 GPa at the machining speeds of 400, 100, and 25 m/s, respectively.

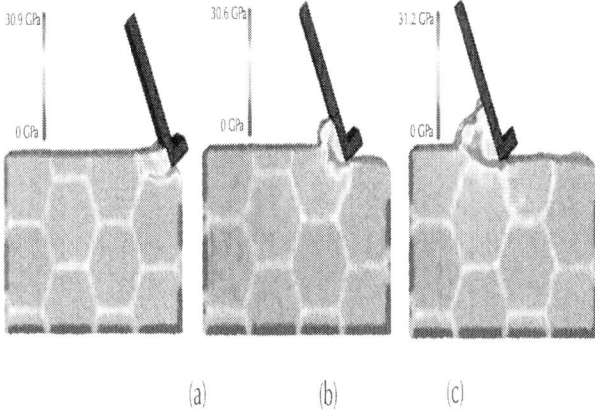

Figure 9: Chip formations and equivalent stress distributions in nano-scale polycrystalline machining for case C8. At the tool travel distances of (a) 30, (b) 120, and (c) 240 Å.

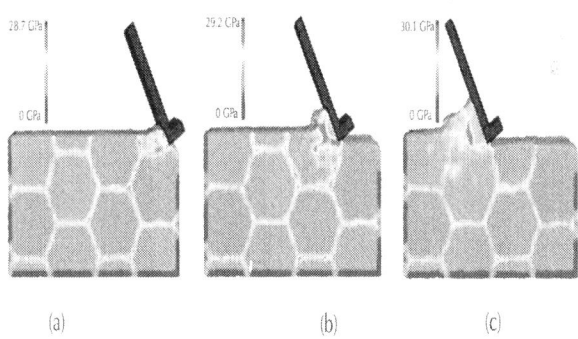

Figure 10: Chip formations and equivalent stress distributions in nano-scale polycrystalline machining for case C9. At the tool travel distances of (a) 30, (b) 120, and (c) 240 Å.

By comparing the cutting force results shown in Figure 11 and Table 6, it is observed that higher machining speeds constantly introduce higher tangential forces, while the increase of thrust force flats out after the machining speed exceeds 100 m/s. Overall, as the machining speed increases from 25 to 400 m/s, the tangential force increases from 339.85 to 412.16 eV/Å and the thrust force increases from 257.03 to 353.59 eV/Å.

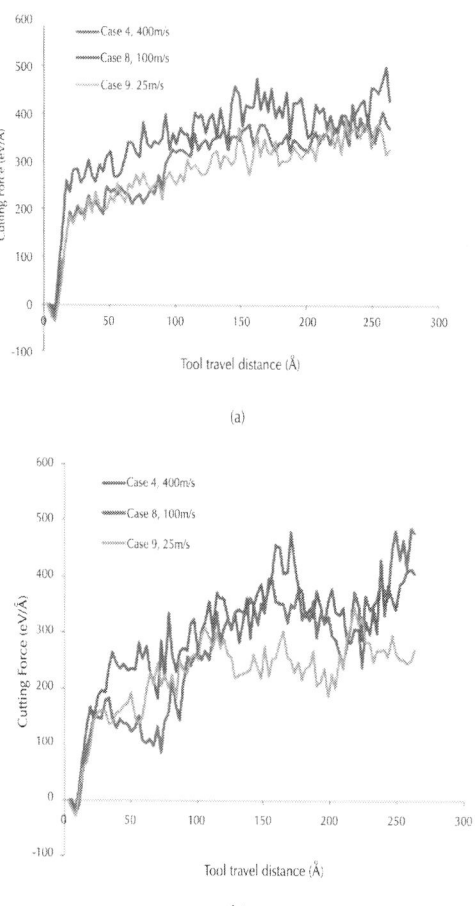

Figure 11: Evolution of cutting forces at the machining speeds of 25, 100, and 400 m/s. (a) Tangential force, F_x and (b) thrust force, F_y.

Table 6: Average cutting force values with respect to machining speed

Case number	Machining speed (m/s)	F_x (eV/Å)	F_y (eV/Å)	F_x/F_y
C4	400	412.16	353.59	1.17
C8	100	358.08	355.02	1.01
C9	25	339.85	257.03	1.32

Shi et al.

Shi et al. Nanoscale Research Letters 2013 8:500, doi:10.1186/1556-276X-8-500

Effect of Grain Size

Cutting Force and Equivalent Stress Distribution

We first investigate the effect of grain size on cutting forces in machining polycrystalline structures. Figure 12 shows the evolution of cutting force components for cases C2 to C7, which represent six polycrystalline structures (i.e., 16.88, 14.75, 13.40, 8.44, 6.70, and 5.32 nm, respectively, in terms of grain size). For benchmarking, the case of monocrystalline machining, namely, case C1, is also added to the comparison. Similarly, the average F_x and F_y values are obtained from the period of tool travel distance of 160 to 280 Å for these cases, and the results are shown in Figures 13 and 14. It is clear that the overall magnitudes of both F_x and F_y for monocrystalline machining are higher than any of the polycrystalline cases. The average F_x and F_y values for case C1 are 470 and 498 eV/Å, respectively. This is reasonable in that the monocrystal copper structure assumes to be perfect without any defects and thus has the highest strength. On the other hand, the existence of grain boundaries, a major form of crystal defects, in all the polycrystalline cases means lower material strengths. Interestingly, the most significant volatility of cutting force is observed in monocrystalline machining. This should be attributed to the highly anisotropic properties of monocrystalline structure and the associated dislocation movement.

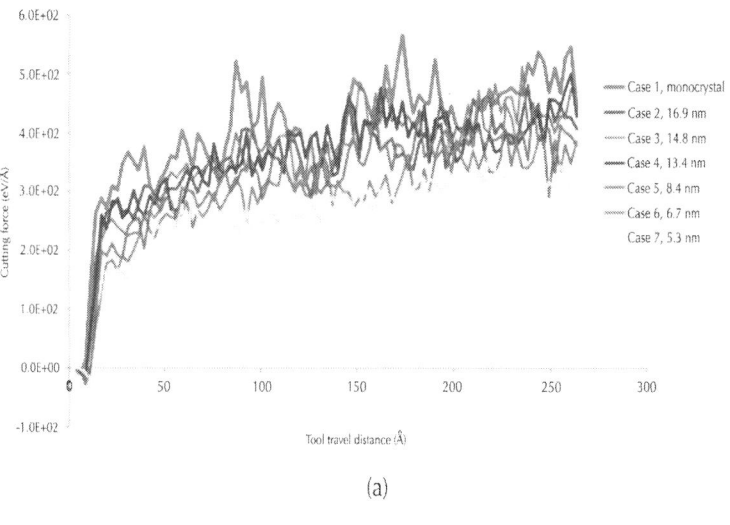

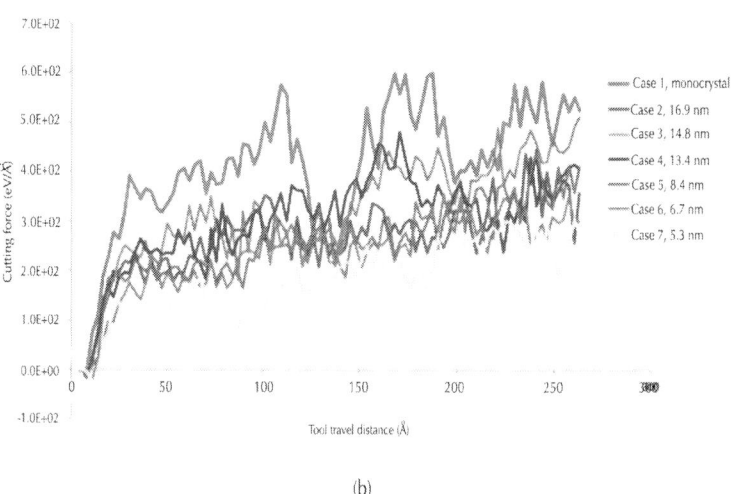

Figure 12: Cutting force evolution in machining polycrystalline coppers of various grain sizes. (a) Tangential force and (b) thrust force.

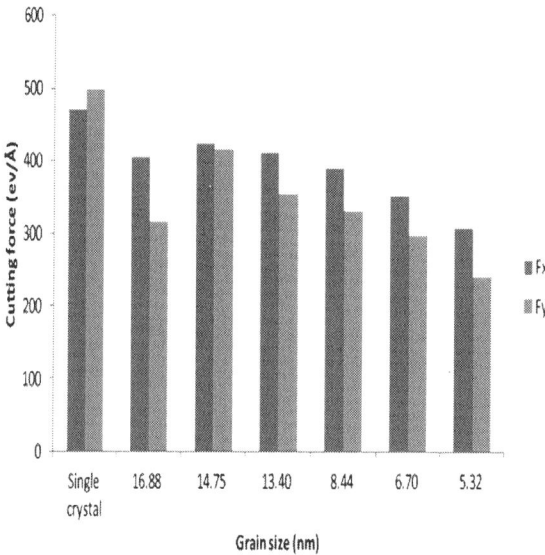

Figure 13: Average tangential and thrust forces for machining polycrystalline coppers of different grain sizes.

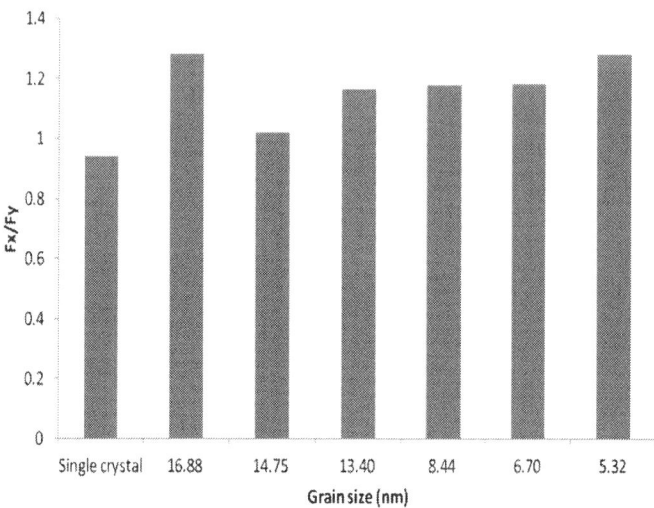

Figure 14: Ratio of F_x/F_y for machining polycrystalline coppers of different grain sizes.

More important observations are made with the six polycrystalline cases. It can be seen from Figure 13 that the average cutting forces increase with the increase of grain size in the range of 5.32 to 14.75 nm. In the range, the relative increases are 37.7% and 72.9% for tangential force and thrust force, respectively. However, the cutting forces reverse the increasing trend when the grain size increases to 16.88 nm (case C7). A similar disruption occurs in the trend of F_x/F_y with respect to grain size, as shown in Figure 14. The ratio of F_x/F_y generally decreases with the increase of grain size, but it rebounds by about 25% when the grain size increases from 14.75 to 16.88 nm. This phenomenon related to grain size and grain boundary is for the first time observed in machining research.

Figure 15 depicts the snapshots (tool travel distance = 240 Å) of equivalent stress distribution for the seven polycrystalline cases with various grain sizes (i.e., cases C1 to C7) at the tool travel distance of 240 Å. For each case, the maximum equivalent stress is found to be in the primary shear zone, and it takes the values of 42.4, 39.5, 42.0, 42.7, 42.5, 41.8, and 41.6 GPa for cases C1 to C7, respectively. It overall agrees with the trend of cutting forces, but the magnitude of stress value change is less drastic.

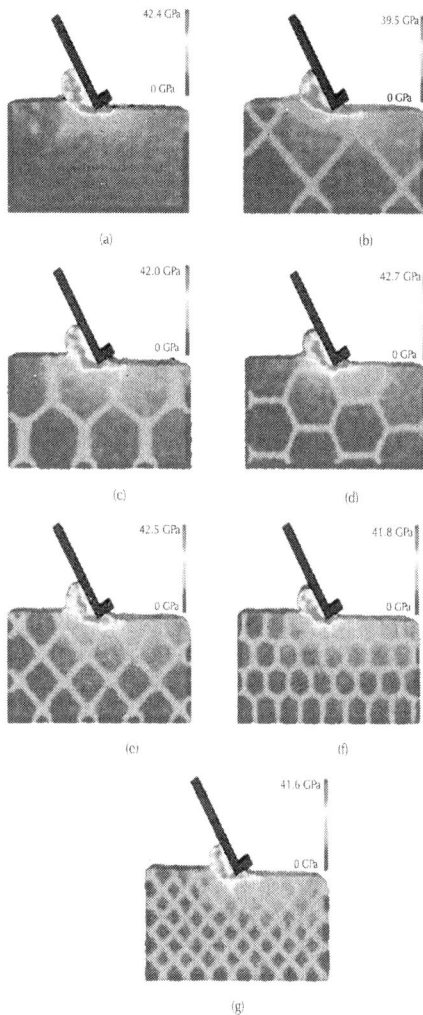

Figure 15: Equivalent stress distributions in machining polycrystalline coppers with different grain sizes. (a) Monocrystal, (b) 16.88 nm, (c) 14.75 nm, (d) 13.40 nm, (e) 8.44 nm, (f) 6.7 nm, and (g) 5.32 nm.

Inverse Hall–Petch Relation

The influence of grain boundary on material properties can be significant, but it depends on the exact conditions of deformation and the particular material used. In the following, we intend to explain the change of cutting forces with respect to grain size in machining polycrystalline coppers. Usually, the strength of polycrystalline materials is expected to increase if the grain size decreases. For coarse-grain materials, the grain size effect on flow stress can be captured by the empirical Hall–Petch relation, which suggests that the yield stress increases with decreasing grain size by the following:

$$\sigma = \sigma_0 + K/\sqrt{d}, \tag{6}$$

where σ_0 is the yield stress, K is the Hall–Petch slope, and d is the grain size. According to this classical relation, the force components in machining polycrystalline copper should increase with the decrease of grain size. Indeed, it is the case when the grain size decreases from 16.88 to 14.75 nm. The tangential force increases by 4.6%, and the thrust force increases by 31.6%. However, the Hall–Petch relation is apparently not applicable for polycrystalline machining with grain sizes of 5.32 to 14.75 nm (i.e., cases C2 to C6), in which the cutting forces decrease with the decrease of grain size.

In recent years, it has been discovered that when the grain size of nano-structured materials is smaller than a critical value, the Hall–Petch relation could be inversed [37-39]. In other words, as the fraction of grain boundary atoms increases to a significant level, work softening will become dominant. The inverse Hall–Petch relation indicates that a smaller grain size increases the volume fraction of grain boundary, which facilitates the activation of other deformation mechanisms such as grain boundary sliding and thereby lowers material strength. The inverse Hall–Petch relation indeed matches up with our observation of nano-scale polycrystalline machining in the particular grain size range. Apparently, the decrease in cutting forces with the decrease of grain size is the result of yield strength reduction. The decrease in cutting force can also be further explained as strengthening due to dislocation activity below a critical grain size is ceased, and the kick-in of other mechanisms leads to work softening and thus lowers the force required by the tool to remove the material.

In particular, Mohammadabadi and Dehghani developed a modified Hall–Petch equation, which incorporates the negative slope observed between grain size and yield stress [40]. It is in the following form:

$$\sigma = \sigma_0 + K/\sqrt{d} - \sigma_{in} f_{gb}, \qquad (7)$$

where σ_{in} is internal stress along the grain boundary that depends on parameters such as grain boundary thickness, lattice distortions, and grain size, and f_{gb} is the volume fraction of the grain boundary. Figure 16 shows the yield stress of polycrystalline copper as a function of grain size under both the conventional Hall–Petch relation and the modified Hall–Petch relation. It can be seen that if the conventional Hall–Petch relation is followed, the yield stress should increase exponentially with grain size reduction. However, the modified Hall–Petch relation indicates that with the decrease of grain size, the yield stress grows at a slower pace to its peak position when the grain size is around 14 nm, and then it starts to drop if the grain size is below this critical value. Note that there are also other literature reporting that for some metals, the critical grain size for the inverse Hall–Petch to take over is about 10 to 15 nm [38, 41-43].

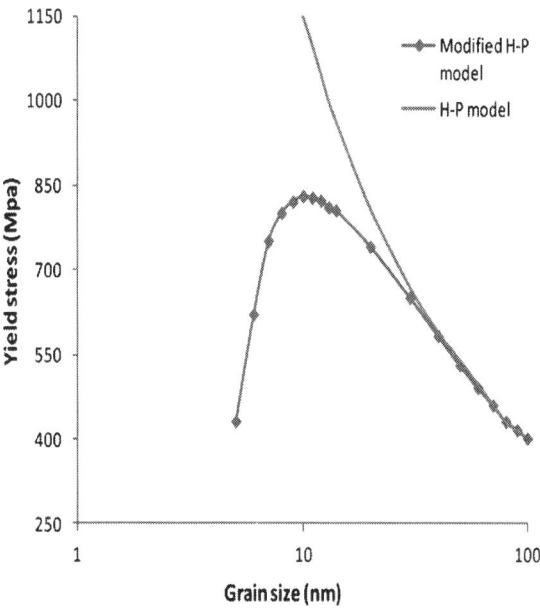

Figure 16: Predicted yield stress for nano-structured copper as a function of grain size. Based on the Hall–Petch and modified Hall–Petch relations [40].

A few other studies also show that the flow stress of ultrafine nanostructured materials can decrease as a result of grain size reduction. With the inverse Hall–Petch effect, the deformation is no longer dominated by dislocation motion, while atomic sliding in grain boundaries starts to play the major role [44]. Narayan experimentally studied this phenomenon by pulsed laser deposition to produce nanocrystalline materials [45]. It was discovered that when the copper nano-crystal is less than 10 nm, material hardness decreases with the decrease of grain size. The decrease in the slope of the Hall–Petch curve and eventually the decrease in hardness below a certain grain size can be explained by a model of grain-boundary sliding [46]. Because of this, as the grain size decreases from 61 to 30 nm, the overall material strength increases, but further decrease in the grain size may result in a decrease of strength. The grain-boundary sliding theory is supported by other researchers [47, 48], where the small and independent slip events in the grain boundary are seen in the uniaxial tension deformation process of fcc metal with a very small grain size (less than 12 nm).

As such, the modified Hall–Petch relation explains well our discoveries in Figure 13. First, the cutting force increase due to the increase of grain size takes place in polycrystalline machining for the grain size range of 5.32 to 14.75 nm. This is in general consistent with the range reported in the literature that the inverse Hall–Petch effect is dominant. Second, the cutting forces decrease when the grain size becomes larger than 14.75 nm. This is exactly where the regular Hall–Petch effect starts to take over. Therefore, in polycrystalline machining, the critical grain size that divides the regular Hall–Petch and inverse Hall–Petch effects is overall consistent with the critical grain size for yield stress in the literature.

It should also be noted that the maximum equivalent stress in our model is always more than an order of magnitude higher than the yield stress presented in the modified Hall–Petch curve in Figure 16. The huge difference can be attributed to two major factors. First of all, the yield stress data in Figure 16 were obtained from experimental measurements on realistic coppers which actually carry extra defects such as voids and substitutes, while the MD simulation assumes perfect crystalline defect-free copper within each grain. In this case, the material strength of the defect-free copper should be much higher. The literature estimates the theoretical yield stress of copper to be within the range of 2 to 10 GPa [49]. More importantly, much higher stresses are observed in MD simulation of machining because of the strain rate effect. It is well known that the flow stress increases with the increase of strain rate [50]. Machining processes always produce extremely high strain rates in the primary and secondary shear zones, much higher than many other manufacturing processes or regular material property tests. For instance, in the case of machining of AISI 1045 steel at 400 m/min, the maximum strain rate is close to 20,000 s^{-1} [34]. On the other hand, the strain rates in material property tests are usually less than 1 s^{-1}. For instance, as the strain rate increases from 10^{-4} to 10^4 s^{-1}, the flow stress of oxygen-free high-conductivity (OCHC) copper increases from 0.8 to 1.5 GPa [51], and the yield stress of tantalum increases from 180 to 700 MPa [52]. Moreover, material flow stress increases even more significantly when the strain rate becomes higher than 10^4 orders of magnitude. Armstrong et al. [53] indicated that the flow stresses of α-Fe at strain rates of 10^4 and 10^6 s^{-1} are 800 MPa and 7GPa, respectively. Swegle and Grady [54] showed that for oxygen-free electronic (OFE) copper, the flow stresses are 200 MPa and 2.8 GPa at strain rates of 10^4 and 10^7 s^{-1}, respectively. The strain rates of the simulated nano-scale machining should be at least 10^8 s^{-1} because it is proportional to machining speed and inversely proportional to chip thickness. This is partially verified by comparing the maximum stress of 43.6 GPa in case C11 (400 m/s machining speed) with that of 30.1 GPa in case C9 (25 m/s machining speed). Based on these two reasons, it is reasonable that the equivalent stress in this MD simulation study is significantly greater than the yield stress shown in the modified Hall–Petch curve.

Grain Boundary and Dislocation Interaction

Figure 17 presents the interaction between grain boundary and dislocation movement inside the work material for the monocrystal case (case C1) and three polycrystalline cases (cases C3, C4, and C7) with a grain size of 14.75, 13.40, and 5.32 nm, respectively. The results are plotted to visualize the changes to the crystalline order of perfect fcc copper. Only defect-related atoms, namely, grain boundary atoms and dislocation atoms, are shown. It can be observed that for the monocrystal copper, the dislocation loops originate from the tool/work interface and/or as-machined surface. The directions of dislocation loops are multiple. It could either propagate along the machining direction beneath the machined surface or penetrate much deeper into the bulk material. Compared with the polycrystalline cases, the dislocation movement in the monocrystal copper is more significant and has greater penetration depth than any of the polycrystalline cases. The cutting force comparison shown above confirms the more drastic dislocation movement that exists in machining monocrystal copper.

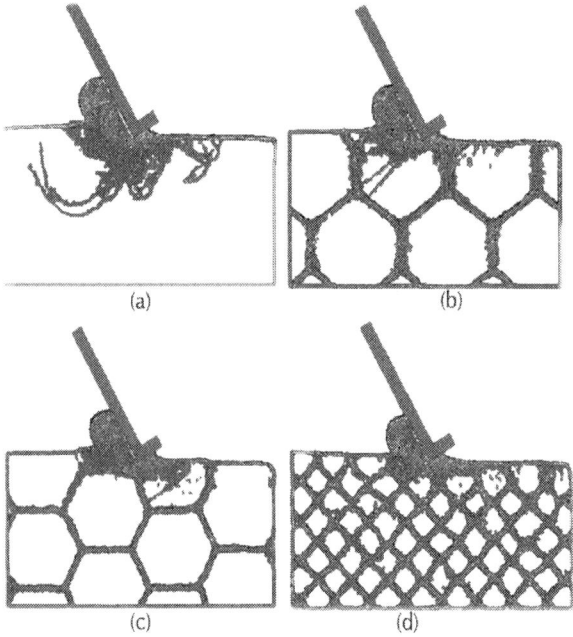

Figure 17: Dislocation development in polycrystalline machining for simulation cases with different grain sizes. (a) Monocrystal, (b) 14.75 nm, (c) 13.40 nm, and (d) 5.32 nm.

As shown in Figure 17b for case C3, since the atomic mismatch between different grains creates a stress field to oppose continued dislocation motion, the dislocations inside grains are clearly blocked by the grain boundary. Therefore, the 'pile-up' phenomenon of dislocation occurs as a cluster of dislocations that are not able to move across the grain boundary. The pile-up phenomenon of dislocations is the hallmark mechanism of the normal Hall–Petch relationship. Due to the resistance effect of the grain boundary to the propagation of dislocation, more force needs be applied to move the dislocations across a grain boundary and hence the increase of yield strength and cutting forces. If the grain size continues to decrease, it falls into the inverse Hall–Petch region, as shown in Figure 17c. In this case, the amount of dislocation movement substantially decreases. This indicates that as the grain size drops below the grain boundary strengthening limit, a smaller grain size would suppress the formation of dislocation pile-ups and instead promotes more grain boundary diffusion and sliding, which resolves the applied stress and in turn reduces the material's yield strength. The grain boundary movement for case C7 can be observed from Figure 17d. The shape of many grains becomes irregular, and the grain boundaries beneath the machined surface slide in response to the exerted cutting forces.

CONCLUSIONS

This paper represents an extensive study of using MD simulation approach to investigate machining of polycrystalline structures at nano-scale. It focuses on two important aspects. One is how machining parameters affect the performance of polycrystalline machining. The other is the influence of grain size of polycrystalline copper structures. For this purpose, we generate 13 simulation cases which cover six levels of grain size, namely, 5.32, 6.70, 8.44, 13.40, 14.75, and 16.88 nm; three levels of machining speed; three levels of depth of cut; and three levels of tool rake angle. The results are analyzed based on cutting forces, stress distribution, chip formation, and dislocation development. The major findings are summarized below:

- Both the tangential and thrust forces increase with the increase of depth of cut for nano-scale polycrystalline machining. The relative increases are 100% and 127% for the tangential and thrust forces, respectively, as the depth of cut increases from 10 to 20 Å. Meanwhile, the maximum equivalent stress value also increases with the depth of cut, but the magnitude of change is much less significant compared with cutting forces.
- Tool rake angle has a significant effect on machining performances in nano-scale polycrystalline machining. As the tool rake angle changes from -30° to +30°, the tangential and thrust forces

decrease by 47% and 1,660%, respectively. The thrust force is much more sensitive to the change of rake angle. The use of nonnegative rake angles reduces the stress concentration in the formed chips.
- The increase of machining speed generally requires higher cutting forces. In the study, the tangential force increases from 339.85 to 412.16 eV/Å and the thrust force increases from 257.03 to 353.59 eV/Å when the machining speed increases from 25 to 400 m/s.
- Thanks to the defect-free lattice structure of monocrystal copper, the cutting forces required are significantly higher for the monocrystalline case compared with all polycrystalline cases investigated.
- Both the regular Hall–Petch relation and the inverse Hall–Petch relation are discovered in investigating the grain size effect in nano-scale polycrystalline machining. In the grain size range of 5.32 to 14.75 nm, the cutting forces increase with the increase of grain size. When the grain size exceeds 14.75 nm, the cutting forces reverse the increasing trend.
- The mechanisms of Hall–Petch and inverse Hall–Petch effects are discussed. The dislocation-grain boundary interaction shows that the resistance of grain boundary to dislocation movement is the fundamental mechanism of the Hall–Petch relation, while grain boundary diffusion and movement is the reason of the inverse Hall–Petch relation.

AUTHORS' CONTRIBUTIONS

Dr. JS conceived of the study and developed the framework of simulation models. Mr. YW carried out the molecular dynamics simulation. Dr. XY provided valuable inputs on the discussion and analysis of results. The first and second authors analyzed the results and drafted the manuscript. All authors read and approved the final manuscript.

ACKNOWLEDGMENTS

The authors would like to thank the valuable inputs from anonymous reviewers for improving the quality of this manuscript.

REFERENCES

1. Inamura T, Takezawa N, Kumakia Y: Mechanics and energy dissipation in nanoscale cutting. *CIRP Ann* 1993, 42(1):79-82.
2. Inamura T, Takezawa N, Kumaki Y, Sata T: On a possible mechanism of shear deformation in nanoscale cutting. *CIRP Ann* 1994, 43(1):47-50.
3. Ikawa N, Shimada S, Tanaka H: Minimum thickness of cut in micromachining. *Nanotechnology* 1992, 3(1):6-9.
4. Fang T, Weng C: Three-dimensional molecular dynamics analysis of processing using a pin tool on the atomic scale. *Nanotechnology* 2000, 11(3):148-153.
5. Shimada S, Ikawa N, Ohmori G, Tanaka H: Molecular dynamics analysis as compared with experimental results of micromachining. *CIRP Ann* 1992, 41(1):117-120.
6. Shimada S, Ikawa N, Tanaka H, Uchikoshi J: Structure of micromachined surface simulated by molecular dynamics analysis. *CIRP Ann* 1994, 43(1):51-54.
7. Ye YY, Biswas R, Morris JR, Bastawros A, Chandra A: Molecular dynamics simulation of nanoscale machining of copper. *Nanotechnology* 2003, 14(3):390-396.
8. Komanduri R, Lee M, Raff LM: The significance of normal rake in oblique machining. *Int J Mach Tool Manuf* 2004, 44(10):1115-1124.
9. Komanduri R, Chandrasekaran N, Raff LM: MD simulation of exit failure in nanometric cutting. *Mater Sci Eng A* 2001, 311(1–2):1-12.
10. Promyoo R, El-Mounayri H, Yang X: Molecular dynamics simulation of nanometric machining under realistic cutting conditions using LAMMPS. In *Proceedings of the ASME 2008*

International Manufacturing Science and Engineering Conference (MSEC2008): October 7–10, 2008; Evanston. New York: ASME; 2008:235-243.

11. Shi J, Shi Y, Liu CR: Evaluation of three dimensional single point turning at atomistic level by molecular dynamics simulation. *Int J Adv Manuf Technol* 2010, 54(1–4):161-171.
12. Zhang L, Tanaka H: Towards a deeper understanding of wear and friction on the atomic scale—a molecular dynamics analysis. *Wear* 1997, 211(1):44-53.
13. Cheng K, Luo X, Ward R, Holt R: Modeling and simulation of the tool wear in nanometric cutting. *Wear* 2003, 255(7–12):1427-1432.
14. Narulkara R, Bukkapatnamb S, Raffc LM, Komanduria R: Graphitization as a precursor to wear of diamond in machining pure iron: a molecular dynamics investigation. *Comput Mater Sci* 2009, 45(2):358-366.
15. Tanaka H, Shimada S, Anthony L: Requirements for ductile-mode machining based on deformation analysis of mono-crystalline silicon by molecular dynamics simulation. *CIRP Ann* 2007, 56(1):53-56.
16. Wang Y, Shi J, Ji C: A numerical study of residual stress induced in machined silicon surfaces by molecular dynamics simulation. *Appl Phys A* 2013. doi:10.1007/s00339-013-7977-8
17. Ji C, Shi J, Wang Y, Liu Z: A numeric investigation of friction behaviors along tool/chip interface in nanometric machining of a single crystal copper structure. *Int J Adv Manuf Technol* 2013, 68:365-374.
18. Lin ZC, Huang JC: A nano-orthogonal cutting model based on a modified molecular dynamics technique. *Nanotechnology* 2004, 15(5):510.
19. Obikawa T, Postek MT, Dornfeld D, Liu CR, Komanduri R, Guo Y, Shi J, Cao J, Zhou J, Yang X, Li X: Micro/nano-technology applications for manufacturing systems and processes. In *Proceedings of the ASME 2009 International Manufacturing Science and Engineering Conference: October 4–7, 2009*. West Lafayette: CD-ROM; 2009.

20. Shi J, Verma M: Comparing atomistic machining of monocrystalline and polycrystalline copper structures. *Mater Manuf Process* 2011, 26(8):1004-1010.
21. Yang B, Vehoff H: Dependence of nanohardness upon indentation size and grain size – a local examination of the interaction between dislocations and grain boundaries. *Acta Mater* 2007, 55(3):849-856.
22. Zhang K, Weertman JR, Eastman JA: The influence of time, temperature, and grain size on indentation creep in high-purity nanocrystalline and ultrafine grain copper. *Appl Phys Let* 2004, 85(22):5197-5199.
23. Li M, Reece MJ: Influence of grain size on the indentation fatigue behavior of alumina. *J Am Ceram Soc* 2000, 83(4):967-970.
24. Lim YY, Chaudhri MM: The influence of grain size on the indentation hardness of high-purity copper and aluminium. *Philosoph Magazine A* 2002, 82(10):2071-2080.
25. Jang H, Farkas D: Interaction of lattice dislocations with a grain boundary during nanoindentation simulation. *Mater Lett* 2007, 61(3):868-871.
26. Zapol P, Sternberg M, Curtiss LA, Frauenheim T, Gruen DM: Tight-binding molecular-dynamics simulation of impurities in ultrananocrystalline diamond grain boundaries. *Phys Rev B* 2001, 65(4):045403.
27. Li J: AtomEye: an efficient atomistic configuration viewer. *Model Simul Mater Sci Engine* 2003, 11:173-177.
28. LAMMPS Molecular Dynamics Simulator http://lammps.sandia.gov/
29. Morse PM: Diatomic molecules according to the wave mechanics. II. Vibrational levels. *Phys Rev* 1929, 34(1):57.
30. Daw MS, Baskes MI: Embedded-atom method: derivation and application to impurities, surfaces, and other defects in metals. *Phys Rev B* 1984, 29(12):6443.
31. Chen H, Hagiwara I, Zhang D, Huang T: Parallel molecular dynamics simulation of nanometric grinding. *Trans Jpn Soc Comput Engine Sci* 2005, 7:207-213.
32. Nieh TG, Wang JG: Hall–Petch relationship in nanocrystalline Ni and Be–B alloys. *Intermetallics* 2005, 13(3–4):377-385.

33. Heino P, Häkkinen H, Kaski K: Molecular-dynamics study of mechanical properties of copper. *Europhys Lett* 1998, 41(3):273.
34. Oxley PLB: *Mechanics of Machining*. Chichester: Ellis Horwood; 1989.
35. Shi J, Liu CR: On predicting chip morphology and phase transformation in hard machining. *Int J Adv Manuf Technol* 2006, 27:645-654.
36. Sreejith PS: Machining force studies on ductile machining of silicon nitride. *J Mater Process Technol* 2005, 169(3):414-417.
37. Lu K, Sui ML: An explanation to the abnormal Hall–Petch relation in nanocrystalline materials. *Scr Metall Mater* 1993, 28(12):1465-1470.
38. Schiøtz J, Jacobsen KW: A maximum in the strength of nanocrystalline copper. *Science* 2003, 301(5638):1357-1359.
39. Koch CC: Optimization of strength and ductility in nanocrystalline and ultrafine grained metals. *Scr Mater* 2003, 49(7):657-662.
40. Mohammadabadi AS, Dehghani K: A new model for inverse Hall–Petch relation of nanocrystalline materials. *J Mater Eng Perform* 2008, 17(5):662-666.
41. Schiøtz J: Atomic-scale modeling of plastic deformation of nanocrystalline copper. *Scr Mater* 2004, 51(8):837-841.
42. Sanders PG, Eastman JA, Weertman JR: Elastic and tensile behavior of nanocrystalline copper and palladium. *Acta Mater* 1997, 10:4019.
43. Schuh CA, Nieh TG: Hardness and abrasion resistance of nanocrystalline nickel alloys near the Hall–Petch breakdown regime. In *MRS Proceedings. Volume 740. No. 1*. Cambridge: Cambridge University Press; 2002. doi:10.1557/PROC-740-I1.8
44. Morris J: The influence of grain size on the mechanical properties of steel. In*Proceedings of the International Symposium on Ultrafine Grained Steels: September 20–22, 2001; Tokyo*. Tokyo: Iron and Steel Institute of Japan; 2001:34-41.
45. Narayan J: Size and interface control of novel nanocrystalline materials using pulsed laser deposition. *J Nanoparticle Res* 2004, 2(1):91-96.

46. Wei YJ, Anand L: Grain-boundary sliding and separation in polycrystalline metals: application to nanocrystalline fcc metals. *J Mecha Phys Sol* 2004, 52(11):2587-2616.
47. Van Swygenhoven H, Derlet PM: Grain-boundary sliding in nanocrystalline fcc metals. *Phys Rev B* 2001, 64(22):224105.
48. Schiøtz J, Di Tolla FD, Jacobsen KW: Softening of nanocrystalline metals at very small grain sizes. *Nature* 1998, 391(6667):561-563.
49. Fan GJ, Choo H, Liaw PK, Lavernia EJ: A model for the inverse Hall–Petch relation of nanocrystalline materials. *Mater Sci Eng A* 2005, 409(1):243-248.
50. Shi J, Liu CR: The influence of material models on finite element simulation of machining. *J Manuf Sci Eng* 2004, 126(4):849-857.
51. Rittel D, Ravichandran G, Lee S: Large strain constitutive behavior of OFHC copper over a wide range of strain rates using the shear compression specimen. *Mech Mater* 2002, 34(10):627-642.
52. Hoge KG, Mukherjee AK: The temperature and strain rate dependence of the flow stress of tantalum. *J Mater Sci* 1977, 12(8):1666-1672.
53. Armstrong RW, Arnold W, Zerilli FJ: Dislocation mechanics of shock-induced plasticity. *Metall Mater Trans A* 2007, 38(11):2605-2610.
54. Swegle JW, Grady DE: Shock viscosity and the prediction of shock wave rise times. *J Appl Phys* 1985, 58(2):692-701.

Chapter 9

Hydraulic Fracturing and its Peculiarities

Stefano Secchi[1] and Bernhard A Schrefler[2]

[1]Institute of system science ISIB – CNR, corso Stati Uniti 4, Padua 35127, Italy

[2]Department of Civil, Environmental and Architectural Engineering, University of Padua, 9 via Marzolo, Padua 35123, Italy

ABSTRACT

Background

Simulation of pressure-induced fracture in two-dimensional (2D) and three-dimensional (3D) fully saturated porous media is presented together with some peculiar features.

Methods

A cohesive fracture model is adopted together with a discrete crack and without predetermined fracture path. The fracture is filled with interface elements which in the 2D case are quadrangular and triangular elements and in the 3D case are either tetrahedral or wedge elements. The Rankine criterion is used for fracture nucleation and advancement. In a 2D setting the fracture follows directly the direction normal to the maximum principal stress while in the 3D case the fracture follows the face of the element around the fracture tip closest to the normal direction of the maximum principal stress at the tip. The procedure requires continuous updating of the mesh around the crack tip to take into account the evolving geometry. The updated mesh is obtained by means of an efficient mesh generator based on Delaunay tessellation. The governing equations are written in the framework of porous media mechanics and are solved numerically in a fully coupled manner.

Results

Numerical examples dealing with well injection (constant inflow) in a geological setting and hydraulic fracture in 2D and 3D concrete dams (increasing pressure) conclude the paper. A counter-example involving thermomecanically driven fracture, also a coupled problem, is included as well.

Conclusions

The examples highlight some peculiar features of hydraulic fracture propagation. In particular the adopted method is able to capture the hints of Self-Organized Criticality featured by hydraulic fracturing.

BACKGROUND

Fluid-driven fracture propagating in porous media is widely used in geomechanics to improve the permeability of reservoirs in oil and gas recovery or of geothermal wells. Another application of importance is related to the overtopping stability analysis of dams. In the case of reservoir engineering, water is forced under high pressure deep into the ground by injection into a well. The fluid, usually mixed with sand and some chemicals, penetrates in the reservoir rock, opening long cracks (fracking). Horizontal drilling together with hydraulic fracturing makes the extraction of tightly bound natural gas from shale formations economically feasible [1]. In the field, it is unfortunately rather difficult to obtain direct information about the evolution of the crack in the ground, and very little data are known or accessible. Two types of measurements are mainly performed: monitoring of pressure fluctuations at the injection pump and registration of acoustic emissions at the surface [2]. Fracking can also induce small earthquakes [3]. Pressure-induced fracture propagation presents some peculiar features such as pressure peaks and stepwise advancement, which have been discovered only recently and need further investigation. It is recalled that differently from tensile experiments where the crack surfaces are stress free, in hydraulic fracturing, these surfaces are loaded by a pressure distribution resulting from the invading fluid or gas [2]. Simulation is an extremely useful tool to obtain more insight into the problem. The paper addresses this issue.

Contributions to the mathematical modelling of fluid-driven fractures have been made continuously since the 1960s, beginning with Perkins and Kern [4]. These authors made various simplifying assumptions, for instance, regarding fluid flow, fracture shape and velocity leakage from the fracture. For other analytical solutions in the frame of linear fracture mechanics, assuming the problem to be stationary, see [5-9]. They suffer the limits of an analytical approach, in particular the inability to represent an evolutionary problem in a domain with a real complexity. An analysis of solid and fluid behaviour near the crack tip can be found in [10,11]. Boone and Ingraffea [12] present a numerical model in the context of linear fracture mechanics which allows for fluid leakage in the medium surrounding the fracture and assumes a moving crack depending on the applied loads and material properties. Tzschichholz and Herrmann [2] used a two-dimensional (2D) lattice model for constant injection rate and homogeneous and heterogeneous material which only breaks under tension. Carter et al. [13] show a fully three-dimensional (3D) hydraulic fracture model which neglects the fluid continuity equation in the medium surrounding the fracture. A discrete fracture approach with remeshing in an unstructured mesh and automatic mesh refinement is used by Schrefler et al. [14]. An element threshold number (number of elements over the cohesive zone) was identified to obtain mesh-independent results. This approach has been extended to 3D situation in [15]. Extended finite elements (XFEM) have been applied to hydraulic fracturing in a partially saturated porous medium by Réthoré et al. [16] in a 2D setting. In this case, a two-scale model has been developed for the fluid flow: in the cohesive crack, Darcy's equation is used for flow in a porous medium, and identities are derived that couple the local momentum and mass balances to the governing equations for the unsaturated medium at macroscopic level. As an example, the rupture of a saturated square plate (0.25 × 0.25 m) in plane strain conditions is investigated under a prescribed fixed vertical velocity $v = 2.35 \times 10^{-2}$ μm/s in the opposite direction at the top and bottom of the plate (tensile loading). The mesh used consists of 20 × 20 quadrilateral elements (12.5 × 12.5 mm each) with bilinear shape functions, and the time step size is 1 s. In the cracked region, the elements are further divided in four triangles. Mohammadnejad and Khoei [17] solve the same problem also with XFEM, using full two-phase flow throughout the region. Darcy flow is assumed within the crack. Finer meshes are used as above (smallest element size 4.5 × 4.5 mm) and much lower time steps (0.25 to 0.125 s). Cavitation is found in both papers, also due to the impervious boundary conditions chosen. Partition of unity finite elements (PUFEM) are used for 2D mode I crack propagation in saturated ionized porous media by Kraaijeveldt et al. [18]. A pull test, a delamination test and an osmopolarity test are simulated with rather fine regular meshes (quadrangular elements with side length of the order of 2 mm and lower) and time step size down to 0.1 s. The time and space discretizations, including the element threshold number used for the solutions, are extremely important for catching the phenomena described next.

We address now the peculiar behaviour of hydraulic fracture propagation which has been observed only by a minority of the above-mentioned authors, but has been confirmed experimentally. Tzschichholz and Herrmann [2] have evidenced with their lattice model and constant injection rate a drop in pressure in time and oscillations on short time scales. These authors explain this by the fact that at the beginning high pressures are needed to push the fluid into the crack. The crack is enlarged and the pressure drops because the enlarged crack can now be opened much more easily than before. The pressure goes down although additional fluid has been added to the crack in the time step. If the pressure drops too much, the stresses at the crack tip fall below their cohesion value and the crack cannot grow at the next time step. By injecting more fluid into the crack, the pressure increases linearly in time until the cohesion forces can be overcome again. Using arguments from continuum mechanics, the authors show that the obtained value for pressure decline in the long term agrees acceptably with their numerical results. The short-term deviations are due the lattice model and the ensuing pressure drops. Oscillations are also obtained for the stored lattice energy. The breaking process is discontinuous in time with time intervals of quiescence where all beams on the crack surface are stressed below their cohesion thresholds and the acting pressure increases linearly in time. Tzschichholz and Herrmann [2] also find a temporal clustering of the breaking events, calling such a sequence bursts (avalanche behaviour). The bursts are unevenly distributed in time and occur relatively often for small times and become rarer later. There is resemblance between the obtained data and magnitude records of earthquakes or of acoustic emission records from laboratory experiments. We have shown with our porous media mechanics model in a 2D setting [14] that in the case of hydraulic fracturing the fracture advances stepwise. Two types of mesh refinement in space and refinement in time were used, but the stepwise advancement did not disappear. Such steps do not appear in other coupled solutions involving cohesive fracture, as e.g. the thermo-elastic one of [19] where the crack surfaces are stress free. The stepwise advancement and flow jumps were also found by Kraaijeveld [20] with a strong and a weak discontinuity model for flow. In [18], the stepwise advancement in mode I crack propagation is difficult to see because a continuous pressure profile across the crack is used. However, continuous pressure profile only works for sufficiently fine meshes. If the mesh is sufficiently fine, then the discretization can resolve the steep pressure gradients along the crack, but the advantage of PUFEM which allows keeping the mesh pretty rough all over the continuum is lost. Hence, dealing with the stepwise progression of the crack in this mode I model is only possible with a finer mesh than the one used (JM Huyghe, personal communication). This is why the authors state that the physical phenomenon challenges the numerical scheme. In mode II, as shown in [20,21], this problem does not appear because a discontinuous pressure across the crack is accounted for. There it is not attempted to resolve the steep pressure gradient, but this gradient is reconstructed afterwards, using the Terzaghi analytical solution for pressure diffusion. This two-step procedure allows using a rough mesh and still handling a realistic pressure gradient. Stepwise crack advancement can clearly be observed in the crack length histories of Figure four of [17], while it does not appear in the solution for the same problem in [16]. The cohesive fracture length for this problem is estimated with Barenblatt's expression (see Equation 22) [22] to be about 136 mm. Hence, there are about 10 elements over the cohesive zone in [16] and 30 elements in [17]. The first value is probably below the element threshold number for this type of problem, even with XFEM (see also the large time steps used), while the second one is sufficient even for standard elements. While both use XFEM, the two-step procedure and the large time step size and coarse mesh in [16] hide the problem.

Finally, stepwise advancement and flow are also mentioned in [23], where PUFEM is used for 2D poroelastic media. Their method still suffers from mesh dependence because the crack propagates through one element each time step. Hence, their conclusions are not definite. However, Pizzocolo et al. [24] confirmed stepwise advancement experimentally with a test on a small hydrogel disk. The duration of the pause Δt between steps is found to be inversely related to the hydraulic permeability K according to $\Delta t = \Delta x^2 / KE$ with E Young's modulus and Δx length of the step. A possible explanation for the stepwise behaviour observed in [20,21,24] put forward in [24] is that an incompressible fluid consolidation comes into play which prevents tip advancement until the overpressures due to the last advancement have been dissipated, and the stress has been transferred again to the solid phase. This implies the existence of pressure peaks after each advancement stage. During the period of quiescence, the effective stress is below the breaking threshold. Consolidation as a possible explanation for the stepwise advancement needs further investigation in the case of fluid injection, because for some permeability values the tip pressure goes down to zero as shown below on an example. The existence of periods of quiescence is in line with the findings of [2]. We will show that this phenomenon is not only relevant for small structures, where it has been observed experimentally, but also for large structures such as underground soil masses and dams. In that case, the fracture length is much larger, but the phenomenon is still there and the bursts can be felt at great distances compared to the crack length.

METHODS

The first subsection presents the fracture model, the second subsection summarizes the governing equations and their numerical solution by means of the finite element method and the third subsection explains the adopted fracture advancement procedure and the required refinements necessary to obtain mesh-independent results.

The Fracture Model

We use a discrete crack model for a situation depicted in Figure 1: Ω is the domain, Γ_e is the boundary of the fully saturated porous material surrounding the crack, Γ is the crack boundary and $\tilde{\Omega}$ is the domain inside the crack filled with fluid only. There is fluid exchange between the crack and the surrounding porous medium. The mechanical behaviour of the solid phase at a distance from the process zone is assumed to obey a Green elastic or hyperelastic material behaviour [14].

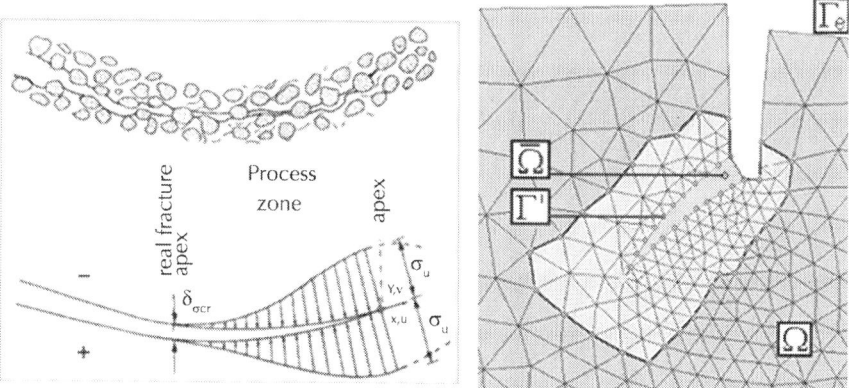

Figure 1: Hydraulic fracture domain and cohesive crack geometry. Definition of the hydraulic fracture domain, reprinted from [15], Copyright (2012), with permission from Springer, and the cohesive crack geometry, reprinted from [14], Copyright (2006), with permission from Elsevier.

For the fracture itself, we use a cohesive fracture model. Between the real fracture apex which appears at macroscopic level and the apex of a fictitious fracture, there is a process zone where cohesive forces act (see Figure 1). Following [22, 25, 26], the cohesive law for mode I crack opening with monotonically increasing opening is σ_0 being the maximum cohesive traction (closed crack), δ_σ the current relative displacement normal to the crack, $\delta_{\sigma cr}$ the maximum opening with exchange of cohesive tractions and $G_c = \sigma_0 \times \delta_{\sigma cr} / 2$ the fracture energy. If after some opening $\delta_{\sigma 1} < \delta_{\sigma cr}$, the crack begins to close and tractions obey a linear unloading as

$$\sigma = \sigma_0 \left(1 - \frac{\delta_\sigma}{\delta_{\sigma cr}}\right) \tag{1}$$

$$\sigma = \sigma_0 \left(1 - \frac{\delta_{\sigma 1}}{\delta_{\sigma cr}}\right) \frac{\delta_\sigma}{\delta_{\sigma 1}} \tag{2}$$

When the crack reopens, Equation 2 is reversed until the opening $\delta\sigma_1$ is recovered; then, tractions obey again Equation 1.

When tangential relative displacements of the sides of a fracture in the process zone cannot be disregarded, mixed mode crack opening takes place. This is often the case of a crack moving along an interface separating two solid components. In fact, whereas the crack path in a homogeneous medium is governed by the principal stress direction, the interface has an orientation that is usually different from the principal stress direction. The mixed cohesive mechanical model involves the simultaneous activation of normal and tangential displacement discontinuity and corresponding tractions. For the pure mode II, the relationship between tangential tractions and displacements is

$$\tau = \tau_0 \frac{\delta_\sigma}{\delta_{\sigma cr}} \frac{\delta_\tau}{|\delta_\tau|} \tag{3}$$

τ_0 being the maximum tangential stress (closed crack), δ_τ the relative displacement parallel to the crack and $\delta_{\sigma cr}$ the limiting value opening for stress transmission. The unloading/loading from/to some opening $\delta_{\sigma 1} < \delta_{\sigma cr}$ follows the same behaviour as for mode I.

For the mixed mode crack propagation, the interaction between the two cohesive mechanisms is treated as in [27]. By defining an equivalent or effective opening displacement δ and the scalar effective traction t as

$$\delta = \sqrt{\beta^2 \delta_T^2 \delta_\sigma^2}, \quad t = \sqrt{\beta^{-2} \tau^2 + \sigma^2} \tag{4}$$

the resulting cohesive law is

$$\mathbf{t} = \frac{t}{\delta}\left(\beta^2 \boldsymbol{\delta}_T + \boldsymbol{\delta}_\sigma\right) \tag{5}$$

β being a suitable material parameter that defines the ratio between the shear and the normal critical components. For more details, see [14].

Governing Equations and their Discretization in Space and Time

Taking into account the cohesive forces and the symbols of Figure 1, the linear momentum balance of the mixture, discretized in space with finite elements according to the standard Galerkin procedure [28] is written as

$$\mathbf{M}\dot{\mathbf{v}} + \int_\Omega \mathbf{B}^T \boldsymbol{\sigma}'' \, d\Omega - \mathbf{Q}\mathbf{p} - \mathbf{f}^{(1)} - \int_{\Gamma'} (\mathbf{N}^u)^T \mathbf{c} \, d\Gamma' = 0 \tag{6}$$

where **c** is the cohesive traction on the process zone as defined above.

The fully saturated medium surrounding the fracture has constant absolute permeability, while for the permeability within the crack, the Poiseuille or cubic law is assumed. This permeability does not depend on the rock type or stress history but is defined by crack aperture only. Deviation from the ideal parallel surface conditions causes only an apparent reduction in flow and can be incorporated into the cubic law, which reads as [29]

$$k_{ij} = \frac{1}{f} \frac{w^2}{12}$$

(7)

w being the fracture aperture and f a coefficient in the range 1.04 to 1.65 depending on the solid material. In the following, this parameter will be assumed as constant and equal to 1.0. Incorporating the Poiseuille law into the weak form of water mass balance equation within the crack and discretizing in space by means of the finite element method results in

$$\tilde{\mathbf{H}}\mathbf{p} + \tilde{\mathbf{S}}\dot{\mathbf{p}} + \int (\mathbf{N}^p)^T \mathbf{q}^w \, d\Gamma' = 0$$

(8)

With

$$\tilde{\mathbf{H}} = \int_{\bar{\Omega}} (\nabla \mathbf{N}^p)^T \frac{w^2}{12\mu_w} \nabla \mathbf{N}^p \, d\bar{\Omega}$$

(9)

$$\tilde{\mathbf{S}} = \int_{\bar{\Omega}} (\mathbf{N}^p)^T \frac{1}{K_f} \mathbf{N}^p \, d\bar{\Omega}$$

(10)

μ_w is the dynamic viscosity and K_f the bulk modulus of the fluid. The last term of (8) represents the leakage flux into the surrounding porous medium across the fracture borders and is of paramount importance in hydraulic fracturing techniques. This term can be represented by means of Darcy's law using the medium permeability and pressure gradient generated by the application of water pressure on the fracture lips. No particular simplifying hypotheses are hence necessary for this term. This equation can be directly assembled at the same stage as the mass balance Equation 11 for the saturated medium surrounding the crack, because both have the same structure: only the parameters have to be changed in the appropriate elements depending whether they belong to the fracture or to the surrounding medium.

The discretized mass balance equation for the porous medium surrounding the fracture is

Hydraulic Fracturing and its Peculiarities

$$\mathbf{Q}^T\dot{\mathbf{u}} + \mathbf{H}\mathbf{p} + \mathbf{S}\dot{\mathbf{p}} - \mathbf{f}^{(2)} - \int_{\Gamma'} \left(\mathbf{N}^p\right)^T \mathbf{q}^w d\Gamma' = 0 \qquad (11)$$

where q^w represents the water leakage flux along the fracture toward the surrounding medium of Equation 7. This term is defined along the entire fracture, i.e. the open part and the process zone. It is worth mentioning that the topology of the domains Ω and $\tilde{\Omega}$ changes with the evolution of the fracture. In particular, the fracture path, the position of the process zone and the cohesive forces are unknown and must be regarded as products of the mechanical analysis.

Discretization in time is then performed with time discontinuous Galerkin approximation following [30,31]. Denoting with $I_n = (t_n^-, t_{n+1}^+)$ a typical incremental time step of size $\Delta t = tn_{+1} - tn$, the weighted residual forms are

$$\int_{I_n} \delta\mathbf{v}^T\left(\mathbf{M}\dot{\mathbf{v}} + \mathbf{K}\mathbf{u} - \mathbf{Q}\mathbf{p} - \mathbf{f}^{(1)}\right)dt + \int_{I_n} \delta\mathbf{u}^T \mathbf{K}(\dot{\mathbf{u}} - \mathbf{v})dt +$$
$$+ \delta\mathbf{u}^T\big|_{t_n} \mathbf{K}\left(\mathbf{u}_n^+ - \mathbf{u}_n^-\right)dt + \delta\mathbf{v}^T\big|_{t_n} \mathbf{M}\left(\mathbf{v}_n^+ - \mathbf{v}_n^-\right) = 0 \qquad (12)$$

$$\int_{I_n} \delta\mathbf{p}^T\left(\mathbf{Q}^T\mathbf{v} + \mathbf{S}\mathbf{s} + \mathbf{H}\mathbf{p} - \mathbf{f}^{(2)}\right)dt + \int_{I_n} \delta\mathbf{p}^T \mathbf{S}(\dot{\mathbf{p}} - \mathbf{s})dt +$$
$$+ \delta\mathbf{p}^T\big|_{t_n} \mathbf{S}\left(\mathbf{p}_n^+ - \mathbf{p}_n^-\right)dt = 0 \qquad (13)$$

with the constraint conditions

$$\dot{\mathbf{u}} - \mathbf{v} = 0$$
$$\dot{\mathbf{p}} - \mathbf{s} = 0 \qquad (14)$$

Subscripts -/+ indicate quantities immediately before and after the generic time station. Field variables and their first time derivatives at time $t \in [t_n, t_{n+1}]$ are interpolated by linear time shape functions, and the following discretized equations are obtained

$$\mathbf{u}_n = \mathbf{u}_n^- + \frac{\Delta t}{2}\left(\mathbf{v}_n^+ - \mathbf{v}_{n+1}^-\right)$$

$$\mathbf{u}_{n+1} = \mathbf{u}_n^- + \frac{\Delta t}{2}\left(\mathbf{v}_n^+ + \mathbf{v}_{n+1}^-\right)$$

$$\mathbf{s}_n = \frac{1}{\Delta t}\left(\mathbf{p}_{n+1} + 3\mathbf{p}_n - 4\mathbf{p}_n^-\right)$$

$$\mathbf{s}_{n+1} = \frac{1}{\Delta t}\left(\mathbf{p}_{n+1} + 3\mathbf{p}_n + 2\mathbf{p}_n^-\right)$$

(15)

$$\left(\frac{1}{2}\mathbf{M} - \frac{5}{36}\Delta t^2 \mathbf{K}\right)\mathbf{v}_n + \left(\frac{1}{2}\mathbf{M} + \frac{1}{36}\Delta t^2 \mathbf{K}\right)\mathbf{v}_{n+1} + \frac{\Delta t}{3}\mathbf{Q}\mathbf{p}_n +$$
$$+ \frac{\Delta t}{6}\mathbf{Q}\mathbf{p}_{n+1} = -\frac{\Delta t}{2}\mathbf{K}\mathbf{u}_n^- + \mathbf{M}\mathbf{v}_n^- + \int_{t_n} N_1(t)\mathbf{f}^{(1)}\,dt$$

$$\left(-\frac{1}{2}\mathbf{M} - \frac{7}{36}\Delta t^2 \mathbf{K}\right)\mathbf{v}_n + \left(\frac{1}{2}\mathbf{M} + \frac{5}{36}\Delta t^2 \mathbf{K}\right)\mathbf{v}_{n+1} + \frac{\Delta t}{3}\mathbf{Q}\mathbf{p}_n +$$
$$+ \frac{\Delta t}{3}\mathbf{Q}\mathbf{p}_{n+1} = -\frac{\Delta t}{2}\mathbf{K}\mathbf{u}_n^- + \int_{t_n} N_2(t)\mathbf{f}^{(1)}\,dt$$

$$\frac{\Delta t}{3}\mathbf{Q}^T\mathbf{v}_n + \frac{\Delta t}{6}\mathbf{Q}^T\mathbf{v}_{n+1} + \left(\frac{1}{2}\mathbf{S} + \frac{\Delta t}{3}\mathbf{H}\right)\mathbf{p}_n + \left(\frac{1}{2}\mathbf{S} + \frac{\Delta t}{6}\mathbf{H}\right)\mathbf{p}_{n+1} =$$
$$= \mathbf{S}\mathbf{p}_n^- + \int_{t_n} N_1(t)\mathbf{f}^{(2)}\,dt$$

$$\frac{\Delta t}{6}\mathbf{Q}^T\mathbf{v}_n + \frac{\Delta t}{3}\mathbf{Q}^T\mathbf{v}_{n+1} + \left(-\frac{1}{2}\mathbf{S} + \Delta t\mathbf{H}\right)\mathbf{p}_n + \left(\frac{1}{2}\mathbf{S} + \frac{\Delta t}{3}\mathbf{H}\right)\mathbf{p}_{n+1} = \int_{t_n} N_1(t)\mathbf{f}^{(2)}\,dt$$

(16)

The nodal displacement, velocity and pressure, $\mathbf{u}_n^-$, $\mathbf{v}_n^-$ and $\mathbf{p}_n^-$, respectively, for the current step coincide with the unknowns at the end of the previous one, hence are known in the time marching scheme and coincide with the initial condition for the first time step. The system of algebraic equations is solved with a monolithic approach using an optimized non-symmetric sparse matrix algorithm. The number of unknowns is doubled with respect to the traditional trapezoidal method.

In a quasistatic situation, adopted for the examples, the submatrices of the above equations are the usual ones of soil consolidation [28], except for

$$\dot{\mathbf{f}}^{(1)} = \int_{\Omega} (\mathbf{N}^u)^T \rho \dot{\mathbf{b}}\, d\Omega + \int_{\Gamma_t} (\mathbf{N}^u)^T \dot{\mathbf{t}}\, d\Gamma + \int_{\Gamma'} (\mathbf{N}^u)^T \dot{\mathbf{c}}\, d\Gamma$$

(17)

where $\dot{c}$ is the cohesive traction rate and is different from zero only if the element has a side on the lips of the fracture Γ'. Given that the liquid phase is continuous over the whole domain, leakage flux along the opened fracture lips is accounted for through the H matrix together with the flux along the crack. Finite elements are in fact present along the crack (not shown in Figure 1), which account only for the pressure field and have no mechanical stiffness. In the present formulation, non-linear terms arise through cohesive forces in the process zone and permeability along the fracture.

Fracture Advancement and Refinement Strategy

Because of the continuous variation of the domain as a consequence of the propagation of the cracks, also the boundary Γ and the related mechanical conditions change. Along the formed crack edges and in the process zone, boundary conditions are the direct result of the field equations, while the mechanical parameters have to be updated. The following remeshing techniques account for all these changes [15, 32].

For the fracture nucleation and advancement, the Rankine criterion is used. More than one crack can open and fractures can also branch. Fracture forms and advances if the maximum principal stress exceeds in a point the fixed threshold. The fracture advancement procedure differs for 2D and 3D situations: in 2D, the fracture follows directly the direction normal to the maximum principal stress, while in 3D, the fracture follows the face of the element around the fracture tip which is closest to the normal direction of the maximum principal stress. In this last case, the fracture tip becomes a curve in space (front). The advancement of a fracture creates new nodes: in 2D, the resulting new elements for the filler at the front are triangles, while in 3D situation, they are tetrahedral. If an internal node along the process zone advances in a 3D setting, a new wedge element results in the filler [15].

At each time station t_n, j successive tip (front) advancements are possible until the Rankine criterion is satisfied (Figure 2). Their number in general depends on the chosen time step increment Δt, the adopted crack length increment Δs and the variation of the applied loads. This requires continuous remeshing with a consequent transfer of nodal vectors from the old to the continuously updated mesh by a suitable operator $v_m(\Omega_{m+1}) = \aleph(v_m(\Omega_m))$. For momentum and energy conservation, the solution is repeated with the quantities of mesh m but re-calculated on the new mesh $m+1$ before advancing the crack tip [33].

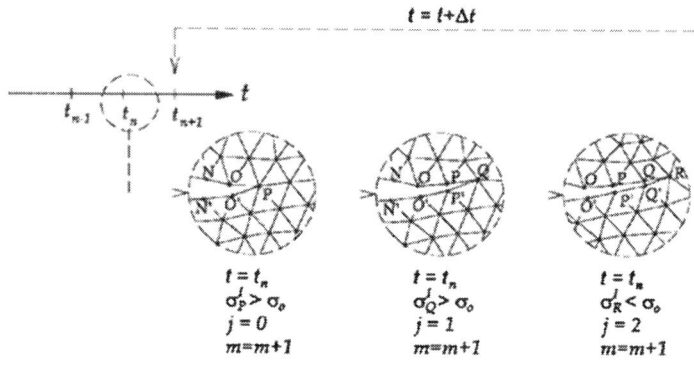

Figure 2: Multiple advancing fracture step at the same time station. Reprinted from [15], Copyright (2012), with permission from Springer.

Three types of refinement are needed to obtain satisfactory results: the refinement in space in general, the satisfaction of an element threshold number over the process zone and a refinement in time. For refinement and de-refinement in space, the Zienkiewicz-Zhu error estimator is used [34]. Fluid lag, i.e. negative fluid pressures at the crack tip, may arise in the case of injection if the speed at which the crack tip advances is sufficiently high so that for a given permeability water cannot flow in fast enough to fill the created space. This as well as mesh-independent results can be obtained numerically only if an *element threshold number* is satisfied over the process zone. This number is given by the ratio of elements over the process zone and its length and can be estimated in advance from the problem at hand and the expected process zone. The number of elements over the process zone is of paramount importance and has not received sufficient attention by many authors. It is a sort of object-oriented refinement and is extensively dealt with in [14]. Adaptivity in time is obtained by means of the adopted discontinous Galerkin method in the time domain (DGT) [33]. The error of the time-integration procedure can be defined through the jump of the solution

$$\langle \mathbf{u}_n \rangle = \mathbf{u}_n - \mathbf{u}_n^-$$
$$\langle \mathbf{v}_n \rangle = \mathbf{v}_n - \mathbf{v}_n^-$$
$$\langle \mathbf{p}_n \rangle = \mathbf{p}_n - \mathbf{p}_n^-$$

(18)

at each time station, i.e. the difference between the final point of time step *n-1* and the first point of time step *n*. By adopting the total energy norms as error measure, we define the following terms:

$$\|\mathbf{e}_u\|_n = \left(\langle \mathbf{v}_n \rangle^T \mathbf{M} \langle \mathbf{v}_n \rangle + \langle \mathbf{u}_n \rangle^T \mathbf{K} \langle \mathbf{u}_n \rangle \right)^{1/2}$$
$$\|\mathbf{e}_{u,p}\|_n = \left(\langle \mathbf{u}_n \rangle^T \mathbf{Q} \langle \mathbf{p}_n \rangle \right)^{1/2}$$
$$\|\mathbf{e}_p\|_n = \left(\langle \mathbf{p}_n \rangle^T \mathbf{Q}^T \langle \mathbf{u}_n \rangle + \langle \mathbf{p}_n \rangle^T \mathbf{H} \langle \mathbf{p}_n \rangle \Delta t + \langle \mathbf{p}_n \rangle^T \mathbf{P}^T \langle \mathbf{p}_n \rangle \right)^{1/2}$$
$$\|\mathbf{e}\|_n = \max \left\{ \|\mathbf{e}_u\|_n, \|\mathbf{e}_{u,p}\|_n, \|\mathbf{e}_p\|_n \right\}$$

(19)

Error measures defined in Equation 19 account at the same time for the cross effects among the different fields and the ones between space and time discretizations.

The relative error is defined as in [30]

$$\eta_n = \frac{\|\mathbf{e}\|_n}{\|\mathbf{e}\|_{max}}$$

(20)

where $\|\mathbf{e}\|_{max}$ is the maximum total energy norm

$$\|\mathbf{e}\|_{max} = \max(\|\mathbf{e}\|_i), 0 < i < n$$

When $\eta > \eta_{toll}$, the time step Δt_n is modified and a new $\Delta t_n' < \Delta t_n$ is obtained according to

$$\Delta t_n' = \left(\frac{\theta \eta_{toll}}{\eta} \right)^{1/3} \Delta t_n$$

(21)

where $\theta < 1.0$ is a safety factor. If the error is smaller than a defined value $\eta_{toll,\,min}$, the step is increased using a rule similar to Equation 21.

As it stands, the refinements in space and time are carried out sequentially, starting with the space refinement, followed by the element threshold number and then the refinement in time. An eye is kept on the satisfaction of the discrete maximum principle [35] which states that it is not possible to refine in time below a certain limit depending on the material properties without also refining in space. A proper functional would be needed to link all the three refinements.

RESULTS AND DISCUSSION

First, we show for comparison purposes the results of cohesive fracture propagation in a thermo-elastic medium, a coupled problem solved with the method for fracture advancement outlined above [19]. The method itself of the crack tip advancement does not introduce steps, once mesh-independent results are achieved. This was also found in static problems (not shown here) such as the three-point bending test, the four-point shear test and the case of a plate with a circular hole. The problem shown here evidences also the importance of the element threshold number, i.e. the number of elements over the process zone. A three-point bending test is performed on a bimaterial specimen subjected to a thermo-mechanical loading [36]. One part of the sample is made of aluminium 6061 and the other of polymethylmethacrylate (PMMA), bonded with methacrylate adhesive. The geometry is presented in Figure 3: the sample has a notch with a sharp tip of 1-mm width and 30-mm height shifted 3 mm from the interface in the PMMA zone. The two materials present very different Young's moduli and thermal expansion coefficients, so that, when the system is subjected to heat, stresses arise near the interface as a result of the mismatch in thermal expansion.

Hydraulic Fracturing and its Peculiarities

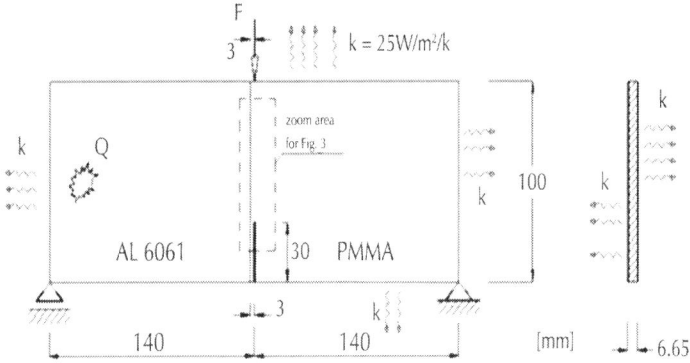

Figure 3: Geometry of the three-point bending test for a bimaterial specimen. Reprinted from [19], Copyright (2004), with permission from Elsevier.

Two different experiments are reported in [36]. In the first, at a room temperature of 25°C, a load was applied 3 mm from the interface in the PMMA zone (Figure 3) to trigger the fracture process. The loading rate was very low and the resulting speed of crack propagation at the initial stages was also quite slow, so that quasistatic conditions can be assumed. The crack path was individuated, and stresses near the crack tip in the PMMA were measured using a shearing interferometer.

In the second experiment, the same operations were performed when the temperature of the aluminium was 60°C in steady state conditions. To reach these conditions, a cartridge heater (Q in Figure 3) was inserted into the aluminium part near the external vertical side. The variation in time of the PMMA temperature was checked before the fracture test, which was performed when steady state conditions were reached. The temperature of PMMA was recorded at the crack tip location, at 5 and 7 mm from the interface. Also in this case, the crack path was spotted. From the differences between the two situations, the authors gathered the thermal effects, which were independent of the magnitude of the applied mechanical load.

In the two experiments, the crack propagation trajectories differ as shown in Figure 4a,b where a zoom of the fractured specimens in correspondence of the notch is presented. In particular, the crack path is closer to the interface when the temperature is higher. The numerical results are shown in Figure 4c. The agreement is remarkable, see also [19].

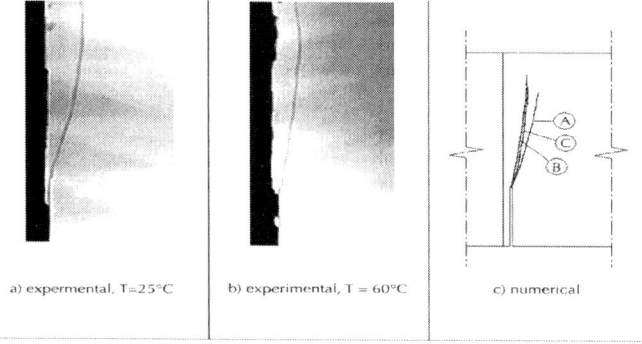

a) expermental, T=25°C   b) experimental, T = 60°C   c) numerical

Figure 4: Zoom of the notch of the specimen with crack path trajectories. (a, b) Experimental results (reproduced from [36]). (c)Numerical results: case A, uniform temperature (25°C); case B, thermal load with E, σ_0, $\delta_{\sigma cr}$ varying with temperature; and case C, thermal load with $E = E(25°C)$, $\sigma_0 = \sigma_0(25°C)$, $\delta_{\sigma cr} = \delta_{\sigma cr}(25°C)$. Reprinted from [19], Copyright (2004), with permission from Elsevier.

Application of Barenblatt's theory [22] for calculation of characteristic cohesive zone size l yields for PMMA

$$\ell = \frac{\pi E G_c}{8(1-v^2)\sigma_0^2} \cong 0.75 \text{ mm}$$

(22)

Our numerical results (0.8 mm) are in good accordance with this value. From this value, the choice of the crack tip advancement length can be estimated. It should be such that the heuristically determined element threshold number is satisfied (five elements in the thermomechanical case). Using linear elements of decreasing size, the value of the force F (Figure 3), corresponding to an applied vertical displacement on the same point, is calculated. Results are summarized in Figure 5. The peak of the external load and the softening branch are mesh independent once the process zone is subdivided into at least five elements with edges of 0.15 mm or smaller. This situation is handled by the mesh generator simply by locating an element source [32] at the crack tip. Its weight may be *a priori* stated and/or can be *a posteriori* updated during the adaptive remeshing procedure once the length of the process zone has been determined. What is important here is that the diagrams in this coupled problem are smooth reasonably once mesh-independent results are obtained.

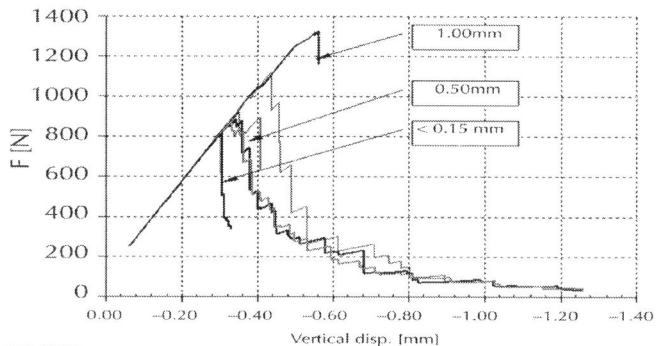

Figure 5: External force vs. vertical displacement and mesh size. Reprinted from [19], Copyright (2004), with permission from Elsevier.

The next application deals with a hydraulically driven fracture due to fluid pumped at constant flow rate Q into a well in 2D conditions (plane strain) [14]. Figure 6 shows the geometry of the problem together with the initial finite element discretization. A notch with a sharp tip is present along the symmetry axis of the analyzed area.

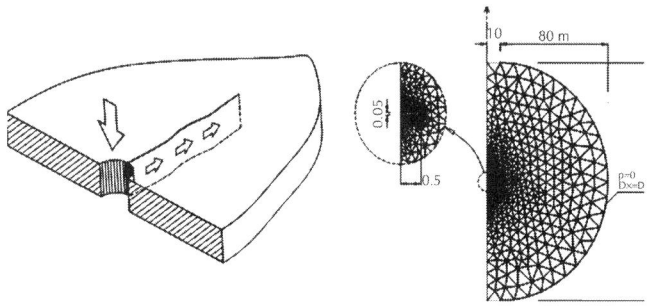

Figure 6: Problem geometry for water injection benchmark and overall discretization. Reprinted from [14], Copyright (2006), with permission from Elsevier.

The effects of combined spatial/temporal discretizations are clearly seen in Figure 7, where the crack length is drawn versus time for different tip advancements, Δs, and time step increments, Δt. The correct time history (case E) is obtained by simultaneously reducing these two parameters, whereas the reduction of only one discretization parameter leads to errors (about ±20%) even using small tip advancement, if compared to the crack length. Again, the importance of the element threshold number is evident for the choice of Δs (the length of the process zone according to Equation 22 is 0.8 m, and about 30 elements are needed over it). It clearly appears that the crack tip velocity is very mesh sensitive. Hence, the element threshold number must be satisfied to obtain mesh-independent results.

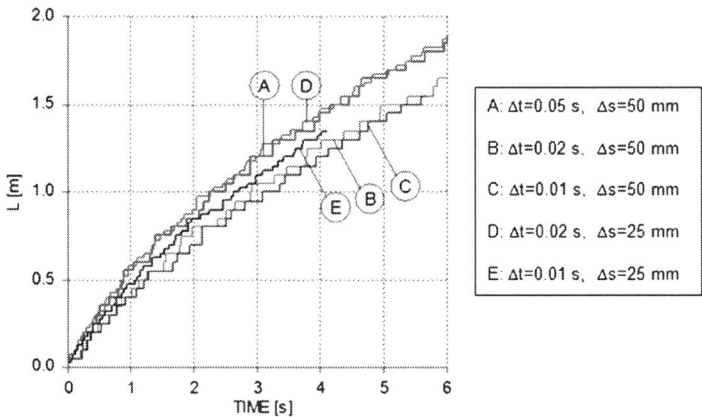

Figure 7: Crack length time history for $\mu_w = 1 \times 10^{-9}$ MPa s and $Q = 0.0001$ m³/s. Reprinted from [14], Copyright (2006), with permission from Elsevier.

A lower number of elements results in wrong crack tip velocity, and the possible development of fluid lag may be missed [14]. Fluid lag corresponds to negative pressure in the process zone and determines hence different body forces (see Equation 6). The distribution of the pressure over the fracture length at time station 10 min is shown in Figure 8 for the following three combinations of dynamic viscosity and injection rate: $\mu_w = 1 \times 10^{-9}$ MPa s, $Q = 0.0001$ m³/s; $\mu_w = 1 \times 10^{-11}$ MPa s, $Q = 0.0001$ m³/s; and $\mu_w = 1 \times 10^{-9}$ MPa s, $Q = 0.0002$ m³/s. The fracture length clearly varies with the chosen data. For the first combination, the pressure at the fracture tip goes almost to zero, while for lower values of μ_w, the pressure is almost constant. For high μ_w and doubled injection rate, cavitation occurs.

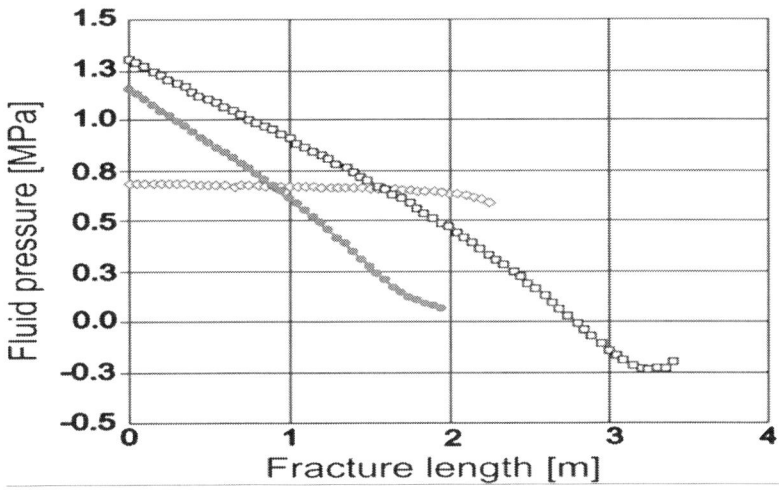

Figure 8: Distribution of the fluid pressure over the fracture length. At time station 10 min for the combinations of dynamic viscosity and injection rate: $\mu_w = 1 \times 10^{-9}$ MPa s, $Q = 0.0001$ m³/s (red circles); $\mu_w = 1 \times 10^{-11}$ MPa s, $Q = 0.0001$ m³/s (green diamonds); and $\mu_w = 1 \times 10^{-9}$ MPa s, $Q = 0.0002$ m³/s (white squares).

The third case deals with the benchmark exercise A2 proposed by ICOLD [37]. The benchmark consists in the evaluation of failure conditions as a consequence of overtopping wave acting on a concrete gravity dam. Contrarily to the previous example here, we have increasing pressure. The geometry of the dam is shown in Figure 9 together with boundary conditions and an intermediate discretization. Differently from the original benchmark, the dam concrete foundation is also considered, which has been assumed homogeneous with the dam body. In such a situation, the crack path is unknown. On the contrary, when a rock foundation is present, the crack naturally develops at the interface between the dam and foundation. In Figure 9 also, the influence of the viscosity on the crack direction is evidenced (circle).

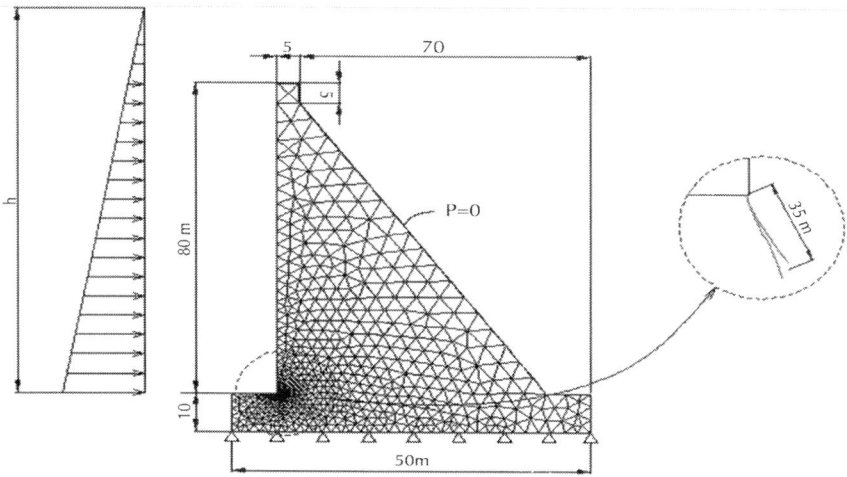

Figure 9: Problem geometry for ICOLD benchmark and calculated crack positions. Reprinted from [14], Copyright (2006), with permission from Elsevier.

The initial condition is obtained under self-weight and the hydrostatic pressure due to water in the reservoir up to a level of 52 m. From this point, the water level increases until the overtopping level is reached (higher than the dam crest [14]). The increase of water level in the reservoir is specified in days according to the benchmark.

For an intermediate situation, the principal stress contours and the cohesive forces are shown on Figure 10. Also, fluid lag has been obtained for this situation, not shown in the picture (see [14]). The crack mouth opening displacement versus days is depicted in Figure 11 for different values of the crack tip advancement. The smallest value corresponds to the proper element threshold number. Clearly, stepwise advancement can be observed with some clustering of the steps.

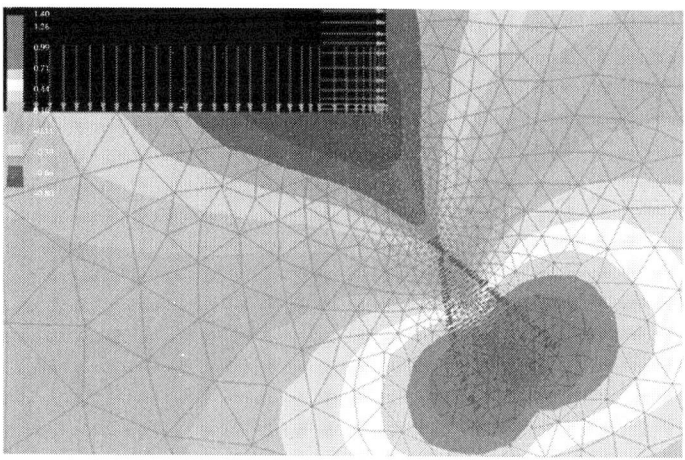

Figure 10: Zoom near the fracture on maximum principal stress contour and cohesive **forces.** Reprinted from [14], Copyright (2006), with permission from Elsevier.

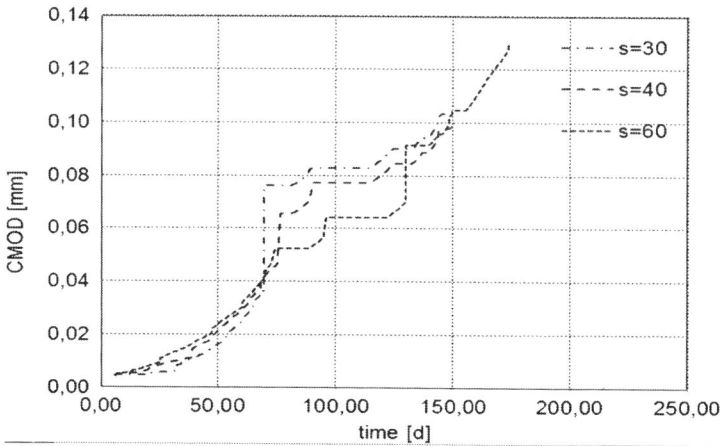

Figure 11: Crack mouth opening displacement versus time (days) for different values of the crack tip advancements (mm).

The effects of the stepwise advancement can also be felt at great distance from the actual crack: the horizontal displacements on the dam crest are effected, as can be seen from Figure 12. Only the diagram for the purely elastic solution (no crack) is smooth. Note that here the vertical scale is logarithmic and in the abscissa appear the time steps, not the actual time. This is the reason why the diagram for the elastic case is above the others.

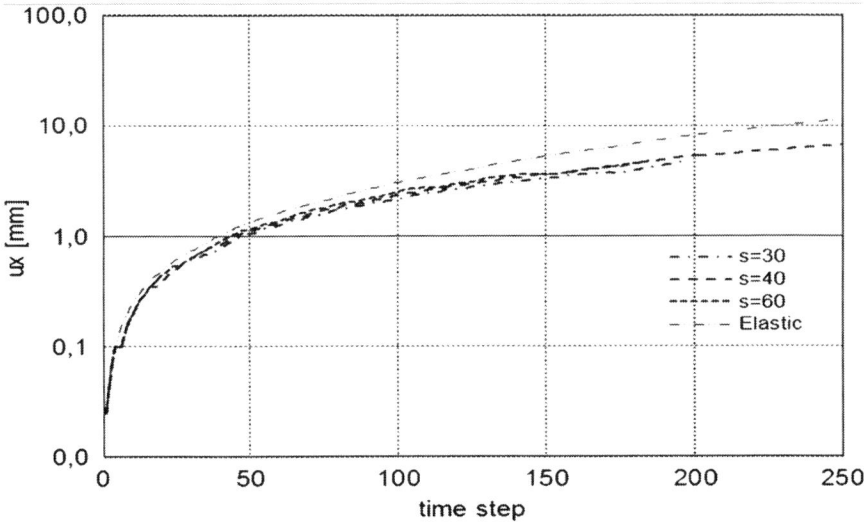

Figure 12: Horizontal displacements versus time step of the dam crest. For different values of the crack tip advancements (mm) and without fracture (elastic).

For a similar problem, a 3D solution has been obtained in [15]. In Figure 13, the mesh, the fracture, the process zone and the stress contours are shown when the fracture length is about 15 m corresponding to an intermediate step of the analysis when the water level is 80 m. The horizontal displacement of the dam crest is drawn versus time in Figure 14. The following situations are considered: no fracture at all (elastic); dry fracture (fracture), i.e. water pressure acts only on the dam, not on the crack lips; hydrostatic water pressure in the crack, constant over the crack length (hydraulic fracture); and fully coupled solution with water pressure varying over the crack length (u-p). The last one has fluid exchange between the crack and surroundings. The results for the last case correspond to an intermediate value between the others because the pressure is diminishing towards the crack tip, reaching even negative values there (cavitation).

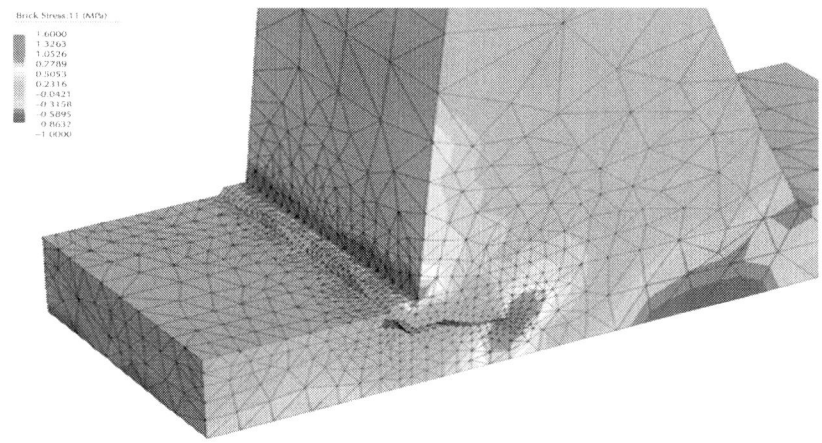

Figure 13: Mesh, fracture, process zone and stress contours. With a fracture length of about 15 m corresponding to an intermediate step of the analysis when the water level is 80 m. Reprinted from [15], Copyright (2012), with permission from Springer.

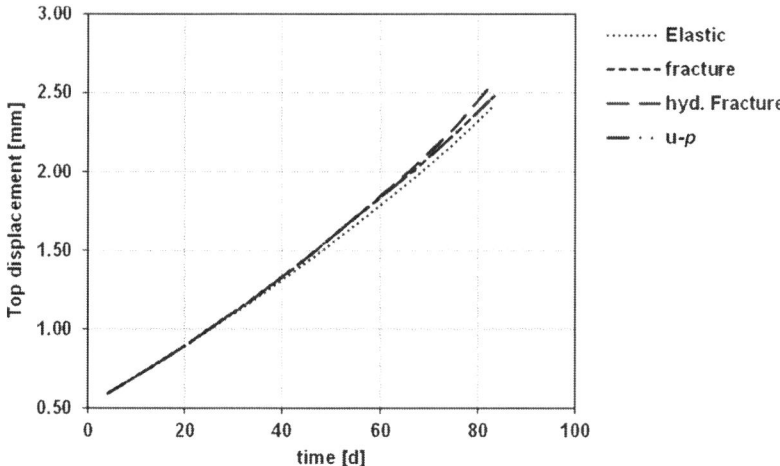

Figure 14: Horizontal displacement of the dam crest versus time. No fracture at all (elastic); dry fracture (fracture), i.e. water pressure acts only on the dam, not on the crack lips; hydrostatic water pressure in the crack, constant over the crack length (hydraulic fracture); and fully coupled solution with water pressure varying over the crack length and fluid exchange between the crack and surroundings (u-p).

The relative variations of the horizontal crest displacements according to

$$\|u\| = \left(\frac{u_i}{u_{el}} - 1\right) \cdot 100 \tag{23}$$

with u_i referring to the studied cases and u_{el} to the elastic solution, are drawn in Figure 15. The largest steps correspond to the situations where fluid is present in the crack and may have pressure exchange (consolidation) with the material surrounding the process zone. Note that these variations are felt on the dam crest, while the pressure-induced fracturing happens on the bottom of the dam. The 3D results have however only qualitative value because the element threshold number would require finer meshes over the cohesive zone which makes the solution very expensive (elements of about 300 mm minimum side length and time steps of a mean value of 4 days were used).

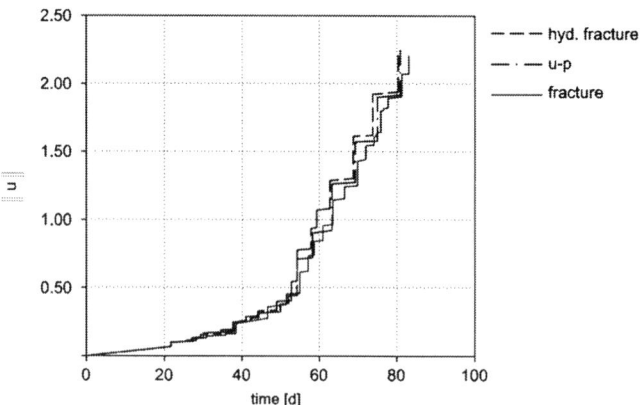

Figure 15: Relative displacements versus time at the dam crest.

In all three examples of hydraulic fracturing, the fracture site is not easily accessible. However, the fact that the effects of the stepwise advancement can be felt also at distance as shown in two examples would make them possible to be monitored remotely. Field data and experimental evidence on reservoir rocks and large bodies are still missing. Data could possibly come from fracking sites or from some fracking-induced earthquakes [3].

CONCLUSIONS

A fully coupled model for pressure-induced cohesive fracture in a saturated porous medium and its solution by the finite element method has been shown. The model is of the discrete crack type and requires continuous updating of the mesh as the crack tip advances. This is achieved with powerful mesh generators. Three types of refinement are necessary to obtain mesh-independent results: a refinement over the domain of the Zienkiewicz-Zhu type, an element threshold number over the process zone and a refinement in time, here with DGT. The results show that in case of pressure induced fracture with pressure exchange and flow between the fracture and the surrounding medium the crack tip advances stepwise. This was found also by few other authors. Smooth diagrams are found on the contrary in a thermo-elastic fracture which is a coupled problem but with stress-free fracture surfaces. From a comparison of results obtained with different methods by other authors, it appears that in some situations a particular adopted method hides the problems discussed in this paper because the required refinements clash with the *raison d'être* of such methods like XFEM or PUFEM (adoption of rough meshes). This is the reason why 'the physical phenomenon challenges the numerical scheme' [18] and why several authors dealing with hydraulic fracturing have not noticed the peculiar behaviour shown here. Also, two-step procedures may introduce some bias in the solution. Two different explanations are found in the literature for the discussed phenomena: one invokes pressure drop [2] and the other pressure peak [24] after crack advancement. The respective loading conditions are different, but the question deserves further scrutiny. Finally, the stepwise advancement may be relevant for earthquake engineering, see e.g. the resemblance between the obtained data of [2] and magnitude records of earthquakes. In many earthquake-prone regions, there is plenty of water available at the level where the rupture takes place [38, 39]. The problem solved in [17] has been solved again with XFEM and finer mesh in [40] and the steps in the fracture advancement featured in [17] disappeared. This implies that XFEM yields a smooth solution for a phenomenon which in nature is not smooth: as shown in [2] hydraulic fracturing exhibits avalanche behaviour and hints of Self-Organized Criticality.

AUTHORS' CONTRIBUTION

SS developed the code, devised the crack tip advancement procedure and carried out the simulations BS drafted the manuscript and explained the results in light of new findings in literature. All authors read and approved the final manuscript.

REFERENCES

1. Vidic RD, Brantley SE, Vandenbossche JM, Joxtheimer D, Abad JD (2013) Impact of shale gas development on regional water quality. Science 340:826-836
2. Tzschichholz F, Herrmann HJ (1995) Simulations of pressure fluctuations and acoustic emission in hydraulic fracturing. Phys Rev E 5:1961-1970
3. Ellsworth WL (2013) Injection induced earthquakes. Science 341:142-150
4. Perkins TK, Kern LR (1961) Widths of hydraulic fractures. SPE J 222:937-949
5. Rice JR, Cleary MP (1976) Some basic stress diffusion solutions for fluid saturated elastic porous media with compressible constituents. Rev Geophs Space Phys 14:227-241
6. Cleary MP (1978) Moving singularities in elasto-diffusive solids with applications to fracture propagation. Int J Solids Struct 14:81-97
7. Huang NC, Russel SG (1985) Hydraulic fracturing of a saturated porous medium—I: general theory. Theor Appl Fract Mech 4:201-213
8. Huang NC, Russel SG (1985) Hydraulic fracturing of a saturated porous medium—II: special cases. Theor Appl Fract Mech 4:215-222
9. Detournay E, Cheng AH (1991) Plane strain analysis of a stationary hydraulic fracture in a poroelastic medium. Int J Solids Struct 27:1645-1662
10. Advani SH, Lee TS, Dean RH, Pak CK, Avasthi JM (1997) Consequences of fluid lag in three-dimensional hydraulic

fracture. Int J Num Anal Methods Geomech 21:229-240
11. Garagash D, Detournay E (2000) The tip region of a fluid-driven fracture in an elastic medium. J Appl Mech 67:183-192
12. Boone TJ, Ingraffea AR (1990) A numerical procedure for simulation of hydraulically driven fracture propagation in poroelastic media. Int J Num Ana Methods Geomech 14:27-47
13. Carter BJ, Desroches J, Ingraffea AR, Wawrzynek PA (2000) Simulating fully 3-D hydraulic fracturing. In: Zaman M, Booker JR, Gioda G (eds) Modeling in geomechanics, Wiley, Chichester. pp 525-567
14. Schrefler BA, Secchi S, Simoni L (2006) On adaptive refinement techniques in multifield problems including cohesive fracture. Comp Methods Appl Mech Engrg 195:444-461
15. Secchi S, Schrefler BA (2012) A method for 3-D hydraulic fracturing simulation. Int J Fracture 178:245-258
16. Réthoré J, de Borst R, Abellan MA (2008) A two-scale model for fluid flow in an unsaturated porous medium with cohesive cracks. Comput Mech 42:227-238
17. Mohammadnejad T, Khoei AR (2013) Hydromechanical modelling of cohesive crack propagation in multiphase porous media using extended finite element method. Int J Numer Anal Meth Geomech 37:1247-1279
18. Kraaijeveldt F, Huyghe JM, Remmers JJC, de Borst R (2013) 2-D mode one crack propagation in saturated ionized porous media using partition of unity finite elements. J Appl Mech 80:020907-1-12
19. Secchi S, Simoni L, Schrefler BA (2004) Cohesive fracture growth in a thermoelastic bi-material medium. Comput Struct 82:1875-1887
20. Kraaijeveld F (2009) Propagating discontinuities in ionized porous media, Dissertation. Eindhoven University of Technology.
21. Kraaijeveldt F, Huyghe JM, Remmers JJC, de Borst R, Baaijens FPT (2014) Shearing in osmoelastic fully saturated media: a mesh-independent model. Engineering Fracture Mechanics.in press
22. Barenblatt GI (1959) The formation of equilibrium cracks during brittle fracture: general ideas and hypotheses: axially-symmetric cracks. J Appl Math Mech 23:622-636

23. Remij EW, Pizzoccolo F, Remmers JJ, Smeulders D, Huyghe JM (2013) Nucleation and mixed-mode crack propagation in porous material. ASCE, Poromechanics V. pp 2260-2269 doi:10.1061/9780784412992.247
24. Pizzocolo F, Hyughe JM, Ito K (2013) Mode I crack propagation in hydrogels is stepwise. Eng Fract Mech 97:72-79
25. Dugdale DS (1960) Yielding of steel sheets containing slits. J Mech Phys Solids 8:100-104
26. Hilleborg A, Modeer M, Petersson PE (1976) Analysis of crack formation and crack growth in concrete by means of fracture mechanics and finite elements. Cem Concr Res 6:773-782
27. Camacho GT, Ortiz M (1996) Computational modelling of impact damage in brittle materials. Int J Solids Struct 33:2899-2938
28. Lewis RW, Schrefler BA (1998) The finite element method in the static and dynamic deformation and consolidation of porous media. Wiley, Chichester.
29. Witherspoon PA, Wang JSY, Iwai KJE, Gale JE (1980) Validity of cubic law for fluid flow in a deformable rock fracture. Water Resour Res 16:1016-1024
30. Li XD, Wiberg NE (1998) Implementation and adaptivity of a space-time finite element method for structural dynamics. Comp Methods Appl Mech Engrg 156:211-229
31. Secchi S, Simoni L, Schrefler BA (2008) Numerical difficulties and computational procedures for thermo-hydro-mechanical coupled problems of saturated porous media. Comput Mech 43:179-189
32. Secchi S, Simoni L (2003) An improved procedure for 2-D unstructured Delaunay mesh generation. Adv Eng Softw 34:217-234
33. Secchi S, Simoni L, Schrefler BA (2007) Numerical procedure for discrete fracture propagation in porous materials. Int J Num Anal Methods Geomech 31:331-345
34. Zhu JZ, Zienkiewic OC (1988) Adaptive techniques in the finite element method. Com Appl Num Methods 4:197-204
35. Rank E, Katz C, Werner H (1983) On the importance of the discrete maximum principle in transient analysis using finite element methods. Int J Num Methods Engng 19:1771-1782

36. Bae JS, Krishnaswamy S (2001) Subinterfacial cracks in bimaterial systems subjected to mechanical and thermal loading. Eng Fract Mech 68:1081-1094
37. ICOLD (1999) Fifth international benchmark workshop on numerical analysis of dams. Theme A2, Denver, Colorado.
38. Doglioni C, Barba S, Carminati E, Riguzzi F (2013) Fault on-off fluids response. Geoscience Frontiers. pp 1-14 doi.org/10.1016/j.gsf.2013.08.004
39. Kelbert A, Schultz A, Egbert G (2009) Global electromagnetic induction constraints on transition-zone water content variations. Nature 460:1003-1006 doi:10.1038/nature08257
40. Mohammadnejad T, Khoei AR (2013) An extended finite element method for hydraulic fracture propagation in deformable porous media with the cohesive crack model. Finite Elements in Analysis and Design 73:77-95

Citations

CHAPTER 1

Mohammad Jamal Khattak and Gilbert Y. Baladi, Analysis of Fatigue and Fracture of Hot Mix Asphalt Mixtures, http://dx.doi.org/10.1155/2013/901652.

CHAPTER 2

Matthew D. Pritzl, Habib Tabatabai, and Al Ghorbanpoor, "Laboratory Assessment of Select Methods of Corrosion Control and Repair in Reinforced Concrete Bridges," International Journal of Corrosion, vol. 2014, Article ID 175094, 11 pages, 2014, doi:10.1155/2014/175094.

CHAPTER 3

Mirosław Witoś, "High Sensitive Methods for Health Monitoring of Compressor Blades and Fatigue Detection," The Scientific World Journal, vol. 2013, Article ID 218460, 31 pages, 2013. doi:10.1155/2013/218460.

CHAPTER 4

S. D. Daxini and J. M. Prajapati, "A Review on Recent Contribution of Meshfree Methods to Structure and Fracture Mechanics Applications," The Scientific World Journal, vol. 2014, Article ID 247172, 13 pages, 2014. doi:10.1155/2014/247172.

CHAPTER 5

Marco Antonio Godoy Jurumenha; João Marciano Laredo dos Reis, Fracture Mechanics of Polymer Mortar Made with Recycled Raw Materials, http://dx.doi.org/10.1590/S1516-14392010000400009.

CHAPTER 6

Lihua Wang, Ze Zhang and Xiaodong Han, In situ M Experimental Mechanics of Nanomaterials at the Atomic Scale, doi:10.1038/am.2012.70.

CHAPTER 7

Goran Ljustina, Martin Fagerström, and Ragnar Larsson, "Rate Sensitive Continuum Damage Models and Mesh Dependence in Finite Element Analyses," The Scientific World Journal, vol. 2014, Article ID 260571, 8 pages, 2014. doi:10.1155/2014/260571.

CHAPTER 8

Jing Shi, Yachao Wang and Xiaoping Yang, Nano-scale machining of polycrystalline coppers - effects of grain size and machining parameters, doi:10.1186/1556-276X-8-500.

CHAPTER 9

Stefano Secchi and Bernhard A Schrefler, Hydraulic Fracturing And Its Peculiarities, doi:10.1186/2196-1166-1-8.

Index

A
Atomic force microscopy (AFM) 167

B
Bulk metal glass (BMG) 26

C
Compacted graphite iron (CGI) 198
Continuum damage model (CD) 215
Crack mouth opening displacement (CMOD) 155

D
Diffuse element method (DEM) 117

E
Element-free Galerkin (EFG) 122
Essential boundary conditions (EBCs) 119, 123
Experimental modal analysis (EMA) 41

F
Finite-element modeling (FEM) 32, 33
Fracture process zone (FPZ) 132

G
Grain boundary (GB) 167, 180

H
High cycle fatigue (HCF) 42
Horizontal plastic deformation (HPD) 2, 3, 5, 22
Hot mix asphalt (HMA) 1
Hot mix asphalt mixtures (HMA) 2

J

Jump Cycle Fatigue (JCF) 99

L

Linear elastic fracture mechanics (LEFM) 132
Linear variable differential transducers (LVDTs) 4
low cycle fatigue (LCF) 42

M

Metal magnetic memory (MMM) 41
Minimum thickness of cut (MTC) 223
Molecular dynamics (MD) 221, 222

P

Polymer mortar (PM) 152
Propulsion Instrumentation Working Group (PIWG) 44

R

Room temperature (RT) 175

S

Scanning electron microscopy (SEM) 28
Splitting Hopkinson pressure bar (SHPB) 26
Stress intensity factors (SIF) 133, 134
Styrene-butadiene-rubber (SBR) 4
Styrene-butadiene-styrene (SBS) 3

T

Thermomechanical fatigue (TMF) 42, 52
Third order shear deformation theory (TSDT) 127
Tip timing (TTM) 41

U

Uncoupled damage model (UD) 215

V

Very high cycle fatigue (VHCF) 42
Voids in mineral aggregates (VMA) 3